Electroweak Symmetry and its Breaking

Electroweak Symmetry and its Breaking

Regina Demina | Aran Garcia-Bellido

University of Rochester, USA

W **World Scientific**

NEW JERSEY · LONDON · SINGAPORE · BEIJING · SHANGHAI · HONG KONG · TAIPEI · CHENNAI · TOKYO

Published by

World Scientific Publishing Co. Pte. Ltd.
5 Toh Tuck Link, Singapore 596224
USA office: 27 Warren Street, Suite 401-402, Hackensack, NJ 07601
UK office: 57 Shelton Street, Covent Garden, London WC2H 9HE

Library of Congress Control Number: 2023933810

British Library Cataloguing-in-Publication Data
A catalogue record for this book is available from the British Library.

ELECTROWEAK SYMMETRY AND ITS BREAKING

ISBN 978-981-122-224-5 (hardcover)
ISBN 978-981-122-225-2 (ebook for institutions)
ISBN 978-981-122-226-9 (ebook for individuals)

For any available supplementary material, please visit
https://www.worldscientific.com/worldscibooks/10.1142/11883#t=suppl

Typeset by Stallion Press
Email: enquiries@stallionpress.com

This book is dedicated to:

Nadejda Mikhailovna Veselova — Thank you for teaching me to use an abacus, this is how it all started;

and

Consuelo Alvarez de Miranda — Thank you for being a constant inspiration to do better.

Preface

This book is geared toward upper level undergraduate students majoring in physics, graduate students specializing in particle physics, as well as experimentalists, who want to brush up on theory, or theorists, who want to have a simple explanation of the experimental methods developed to discover and study the Higgs boson. We expect familiarity with quantum mechanics and the basics of special relativity. It is best to have a course on quantum electrodynamics (QED) prior to reading this book. Knowing some elements of group theory will also be helpful, though not required. If you feel rusty in these areas, we include a very short (cheat-sheet level) review of these subjects in the Appendix. Throughout the text we add *Boxes*, where we include a historic interlude, a more detailed description of an experiment or a tangential discussion. If in a rush, they could be skipped without losing the train of thought.

Contents

Chapter 1

Introduction

Electricity is a well-known force of Nature to anyone who plugged a tea kettle into an electric outlet or paid the electric bill. There is no bill for the weak force though maybe there should be, since weak interactions are responsible for the solar energy. J. C. Maxwell (Fig. 1.1) showed that electric and magnetic forces are the two sides of the same coin. In other words, they can be combined into one force: electromagnetism. The unification of fundamental interactions proved to be very addictive. Since then physicists have been able to add the weak nuclear force to the mix, and continue searching for the unification of all fundamental forces, which include also the strong nuclear force and gravity.

The roots of the combined electroweak theory go deep into the past. Oskar Klein was the first person to suggest the concept of the local isotopic invariance, the cornerstone principle of this theory, in the late 1920s. He did it using five-dimensional space. Wolfgang Pauli (Fig. 1.2 left) suggested the existence of the neutrino, a particle that participates exclusively in weak interactions, to explain apparent energy non-conservation in beta-decays in the 1930s. Enrico Fermi (Fig. 1.2 right) introduced four-fermion interactions to build the first theory of the weak force in the 1940s.

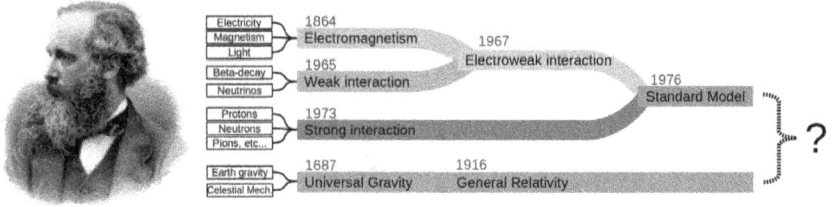

Figure 1.1: James Clerk Maxwell (left) and the progress in the unification of fundamental forces (right).

Figure 1.2: Wolfgang Pauli (left) and Enrico Fermi (right).

Finally, Ginzburg and Landau (Fig. 1.3) suggested a 4th power potential to explain the second-order phase transitions in 1950s. This mechanism was largely adopted by Peter Higgs, (and five other people) to build the model of the electroweak symmetry breaking in 1964. One of the predictions of this model was the existence of a fundamental scalar boson, later named after Peter Higgs. Some 49 years later, on July 4, 2012, the discovery of this particle was announced by CERN, the European Organization for Nuclear Research (Fig. 1.4). The remarkable part was not that it

Figure 1.3: Vitaly Ginzburg and Lev Landau.

Figure 1.4: From left to right, F. Englert, P. Higgs, C. Hagen, and G. Guralnik at the press conference at CERN on July 4, 2012.

was a surprise, quite the opposite. Should the particle fail to reveal its presence, a press release was in the making to explain why that result would be truly revolutionary. And revolutionary it would have been. Yet, just like your car keys are found in the last pocket you check, the Higgs boson was found in the last sliver of parameter space available to it. Since then, a number of measurements of the properties of this particle have been performed, confirming its identity and the theoretical predictions. The existence of this particle confirms the mechanism that is responsible for breaking the symmetry between the weak and electromagnetic interactions. As one of the parents of this mechanism, Carl Hagen, said: "We made the simplest possible assumptions. It is truly surprising that Nature seems to have followed the same logic".

CERN is preparing a massive upgrade of the accelerator and detector complex to significantly increase the size of the available dataset containing Higgs bosons. The physics community is actively engaged in the discussions about building a dedicated Higgs factory. All these with the single goal of deepening our understanding of this particle and its role in the world of fundamental interactions.

In this book, we discuss the theory of the weak force, its unification with the electromagnetic force, in other words the symmetry between the two forces, and the breaking of this symmetry. The construction of the model of electroweak interactions went through a very intricate interplay between theory and experiment. In this book, we do not take the historic approach to the description of the model. Rather we describe the contemporary, essentially complete theory of unified electroweak interactions and refer to experiments that manifest a particular property of the model. After all, unlike mathematics, physics strives to describe the world that we live in, not one of the worlds that might have existed.

Throughout the book we use natural units with Planck constant and the speed of light equal to one: $\hbar = c = 1$. Electric charge is measured in the units of the elementary charge e. We shall use Greek indices to label four-dimensional coordinates, with 0 being the time-like component and $1, 2, 3$ being the space-like components. Latin indices are used to label three-dimensional coordinates. We use the Einstein convention, where summation is implied over the repeating indices.

1.1 Symmetries

Generally, humans associate beauty with symmetry. It is probably even more true for physicists. Few experiences produce such a nice, warm, almost sensual feeling as when a symmetry is introduced. Smear a shapeless blob of paint on a sheet of paper. No beauty is in sight. Fold the paper in two and let the paint smear to the other side. Now that a symmetry is introduced immediately the image becomes more appealing. In this case we are dealing with mirror, aka axial symmetry. There is an axis (in this case the fold in the paper sheet) around which the image is reflected. Reflection is the transformation of this symmetry. The original and the transformed images are identical. Symmetry can be defined as a *non-observability* of some quantity. If you find yourself in the *mirror land* it will take you a while to figure this out. (Incidentally, an experiment involving weak interaction can help you answer this question.) Let us make another mental (gedanken) experiment. Suppose you live in a very boring universe consisting of just the Sun and the Earth, both of negligible size. Let us define the spherical coordinate system (r, θ, ϕ) with its origin on the Sun. The strength of the gravitational interaction depends on the distance between the two bodies r, but not the angles (θ, ϕ). If overnight the Earth is moved to the other side of the Sun, or to any place on a sphere of radius r, centered on the Sun, there is no experiment that you can perform to figure this out. The non-dependence on the angles characterizes the spherical symmetry. Rotation is the *transformation* of this symmetry. In terms of theoretical physics, the Lagrangian describing this system should not depend on angles (θ, ϕ). If it does — the symmetry is broken.

Another profound consequence of the fact that the gravitational force exhibits this symmetry under rotations is that there has to be an associated quantity that is conserved (it remains constant over time). It is a theorem formulated by Emmy Noether (Fig. 1.5): "If the Lagrangian of a system has a continuous symmetry, then there exists an associated quantity which is conserved by the system, and

Figure 1.5: Emmy Noether, probably around 1915.

vice versa." In our example of the Earth–Sun system and the rotational symmetry, it is the angular momentum that is conserved. The proof and a simple example can be found in Section A.4.1. Table 1.1 shows some common examples of non-observables, the symmetry transformations and their conserved quantities. Noether's theorem is an incredibly powerful insight! The origin of the conservation of angular momentum (or energy) in a system emanates from the invariance of its physical description under a rotational transformation (or time translation). This is an important tool for physicists who can now test if a variable is conserved and then learn that the description of the system must be symmetric under certain transformations.

In mathematics, symmetry transformations frequently compose a *group*. Unitary transformations do not change the length of a vector. If any two members of the group commute with each other $(AB = BA)$, the group is called Abelian, otherwise it is non-Abelian. Both cases play an important role in physics. For example,

Table 1.1: Summary of common symmetries, their associated conserved quantities and the non-observables they are associated with.

Non-observable	Symmetry transformation	Conservation law or selection rule
Absolute spatial position	Space translation $\vec{r} \to \vec{r} + \vec{d}$	Momentum
Absolute time	Time translation $t \to t + \tau$	Energy
Absolute spatial direction	Rotation $\hat{r} \to \hat{r}'$	Angular momentum
Absolute left or right	$\vec{r} \to -\vec{r}$	Parity
Absolute sign of electric charge	$e \to -e$ (or $\psi \to e^{i\alpha}\psi^{\dagger}$)	Charge conjugation
Relative phase between different normal states	Gauge transformation $\psi \to e^{iN\theta}\psi$	Particle number: charge, baryon or lepton number
Difference between different coherent mixture of p and n states	$\left(\frac{p}{n}\right) \to u\left(\frac{p}{n}\right)$	Isospin
Difference between identical particles	Permutation (particle substitution)	Bose or Fermi statistics

electromagnetic force is described by an Abelian group U(1), while weak interactions are described by non-Abelian group SU(2). Here the U refers to unitary matrices and SU to the subgroup of "special unitary" matrices with determinant 1. So, let us see where the road of symmetry leads us. (Spoiler alert: it leads to symmetry breaking.)

1.2 The Cast

Before we begin our journey we need to introduce the cast of players, as shown in Fig. 1.6. In the standard model of particle physics,

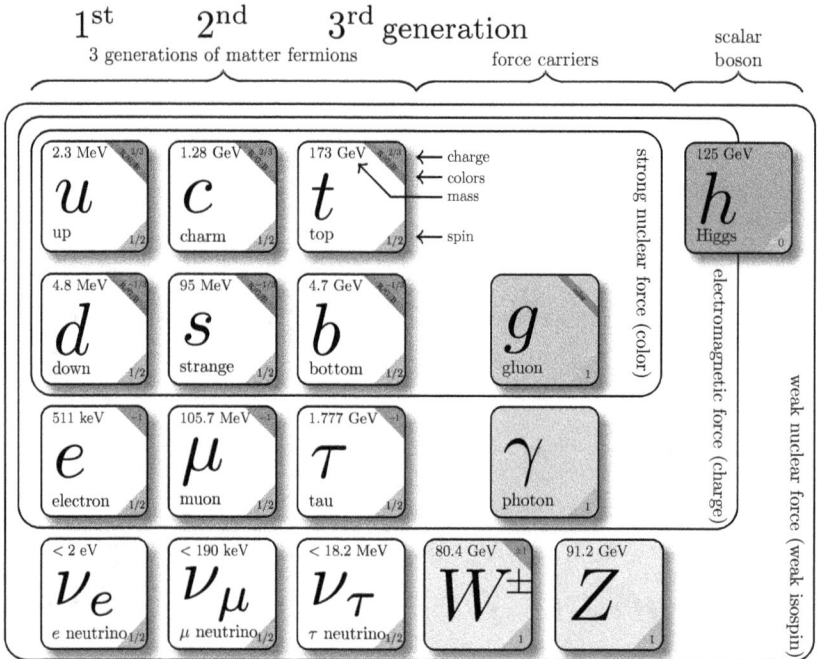

Figure 1.6: The particle content of the standard model: fermions, bosons and how they interact.

the building blocks of the universe are fermions. In general, any particle with half integer spin follows the Pauli–Fermi statistics and is referred as a fermion. In practice, all elementary, i.e., unbreakable, fermions have spin 1/2. There are two types of fermions: quarks, which participate in strong nuclear interactions, and leptons, which do not. Both quarks and leptons are arranged in pairs, a.k.a doublets, and exist in three generations (there are good reasons, which will be discussed later, to believe that there are only three of them). The first generation leptons are an electron e^-, which has a negative electric charge -1 and an electron type neutrino ν_e, which is electrically neutral. The first generation quarks are *up* quark u with electric charge of $+2/3$ and *down* quark d with electric charge of $-1/3$. These properties are reincarnated in the second generation represented by a muon μ and muon type neutrino ν_μ and *charm* quark c and *strange* quark s. The third generation consists of the τ lepton and τ type neutrino ν_τ and *top* quark t and *bottom* quark b. Fermions of each subsequent generation are heavier and unstable, i.e., they typically decay into particles of the lower generation. All atoms of the entire visible universe are made up of the elementary particles of the first generation. The combination of two *up* and one *down* quarks constitutes a charge $+1$ particle, which is a proton. Neutral neutrons consist of two *down* and one *up* quarks. For each fermion there exists an anti-fermion with the opposite electric charge.

The energy sector of the standard model is represented by the mediators of the fundamental interactions. Being of integer spin, they obey Bose–Einstein statistics and are referred to as bosons. Electromagnetic interactions are mediated by photons γ, which are the quanta of the electromagnetic field. They have no electric charge. Weak interactions are mediated by charged W and neutral Z bosons, which are massive ($m_W = 80$ GeV and $m_Z = 91$ GeV). Strong interactions are mediated by eight massless gluons. All of these have spin 1. The hypothetical mediator of gravity is referred to as graviton and is believed to have spin 2 and be massless. Finally, a very special particle needs to be introduced, the Higgs boson h. It has zero spin, hence it is referred to as a scalar field. It is not a force, it is rather a generator of inertia. This is the star of the show, most of the book is dedicated to the discussion of its properties.

Suggested Reading for Chapter 1

[1] Peter W. Higgs. "Broken Symmetries and the Masses of Gauge Bosons". *Phys. Rev. Lett.* 13 (1964), pp. 508–509.

[2] F. Englert and R. Brout. "Broken Symmetry and the Mass of Gauge Vector Mesons". *Phys. Rev. Lett.* 13 (1964), pp. 321–323.

[3] G. S. Guralnik, C. R. Hagen, and T. W. B. Kibble. "Global Conservation Laws and Massless Particles". *Phys. Rev. Lett.* 13 (1964), pp. 585–587.

[4] T. D. Lee and Chen-Ning Yang. "Question of Parity Conservation in Weak Interactions". *Phys. Rev.* 104 (1956), pp. 254–258.

[5] Emmy Noether. "Invariant Variation Problems". *Gott. Nachr.* 1918 (1918), pp. 235–257. arXiv: physics/0503066.

[6] Oskar Klein. "Quantum Theory and Five-Dimensional Theory of Relativity. (In German and English)". *Z. Phys.* 37 (1926), pp. 895–906.

[7] E. Fermi. "Trends to a Theory of beta Radiation. (In Italian)". *Nuovo Cim.* 11 (1934), pp. 1–19.

[8] V. L. Ginzburg and L. D. Landau. "On the Theory of Superconductivity". *Zh. Eksp. Teor. Fiz.* 20 (1950), pp. 1064–1082.

Chapter 2

Electroweak Interactions

It was observed in multiple nuclear scattering experiments that charged and neutral pions behave very similarly as far as strong nuclear interactions are concerned, yet their lifetimes differ enormously. Neutral pions decay into a pair of photons in a matter of 10^{-16} seconds, while charged pions decay into a muon and a neutrino after about 2×10^{-8} seconds, which is a difference of about eight orders of magnitude. If in the first case it is like having a cup of coffee, in the second — millennia pass by, kings replace one another, crusades start and fail... Clearly different physics is at play. While the decay of neutral pions is electromagnetic in nature (since it involves photons), the decay involving charged pions is mediated by a different force, called weak nuclear force (Fig. 2.1). The characteristic feature of this force is, well, its weakness compared to the electromagnetic interactions as demonstrated by the difference in charged and neutral pion lifetimes.

It is worth noting that while electromagnetic interactions do not transfer the charge of the interacting particles, weak interactions described above do (in the right diagram in Fig. 2.1, the total positive charge of +1 is passed on from the initial to the final state). For this reason they are referred to as charged currents.

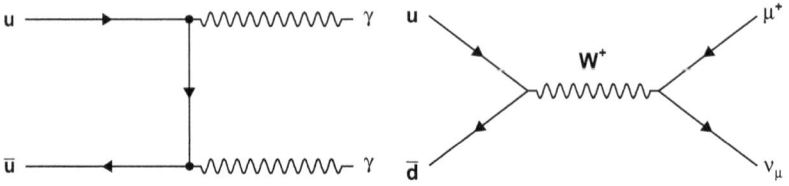

Figure 2.1: The neutral pion $(u\bar{u})$ decay is mediated by the electromagnetic interaction (left plot). The charged pion $(u\bar{d})$ decay, in contrast, is mediated by the weak interaction (right plot), which makes the lifetime of the charged pion several orders of magnitude different to the neutral pion (see Section 2.2).

As we shall soon see, there are also weak interactions that do not transfer the charge, and are thus referred to as neutral currents.

2.1 Helicity, Chirality and P-Parity

There is another notable difference of weak interactions with electricity: they do not conserve P-parity. Let us explain what this means. In our discussion of symmetry we introduced an axial symmetry: the identity between an object and its mirror reflection. If for all objects we ascribe a number called P-parity that is equal to 1, the mirror reflection will have P-parity equal to -1.

A right coordinate system appears as left in the mirror. Thus, mirror reflection is identical to transformation from right-handed coordinate system to a left-handed one, or vice versa. Under this transformation, true vectors $\vec{v}$ are transformed to minus themselves, $-\vec{v}$. Examples of such vectors are a particle's location $\vec{r}$, its momentum $\vec{p}$ and electric field $\vec{E}$. If a vector can be presented as a cross product of two true vectors, it is clear that under P-transformation such a vector would transform into itself. Such vectors are called axial vectors. Examples are angular momentum $\vec{L}$ and magnetic field $\vec{B}$. A scalar product of two vectors or two axial vectors is a true scalar that is transformed into itself under

Figure 2.2: Chen-Ning Yang (left) and Tsung-Dao Lee.

P-transformation. A product of a true vector and an axial vector transforms into minus itself and is referred to as an axial. The Lagrangian describing electricity (see Eq. (A.60)) is invariant under P-transformation. This is not true for the weak force, as was first suggested by Tsung-Dao Lee and Chen-Ning Yang (see Fig. 2.2) based on the analysis of the available experimental data. They found the appropriate description, and finally their predictions were confirmed experimentally by Mdm. Wu (see Box 2.1).

Box 2.1 Experimental Observation of P-Parity non-Conservation by Chien–Shiung Wu (1957)

Madam Wu set out to test the parity properties of neutron decay and to check whether the weak force (unlike electromagnetism or the strong force) may violate parity, as hypothesized by Yang and Lee.

(Continued)

(*Continued*)

She used the beta decay of Cobalt-60 by aligning its nuclear spin with a magnetic field: $^{60}_{27}Co \rightarrow \, ^{60}_{28}Ni + e^- + \bar{\nu} + 2\gamma$.

Let's apply the P-parity transformation to the whole experiment: the magnetic field and the nuclear spin do not change signs under parity transformation (they are axial vectors), the momenta of the electron and the antineutrino change their signs because they are true vectors. Thus, the experimental set-up (B-field and nuclear spins) remains unchanged under parity, but the final configuration (what we observe) is flipped because electrons and neutrinos would be emitted in the opposite directions. Therefore, if parity is conserved, we should observe with equal likelihood the electron parallel to and antiparallel to the direction of nuclear spin. If parity is not conserved, then we will see more electrons in one direction than the other.

(*Continued*)

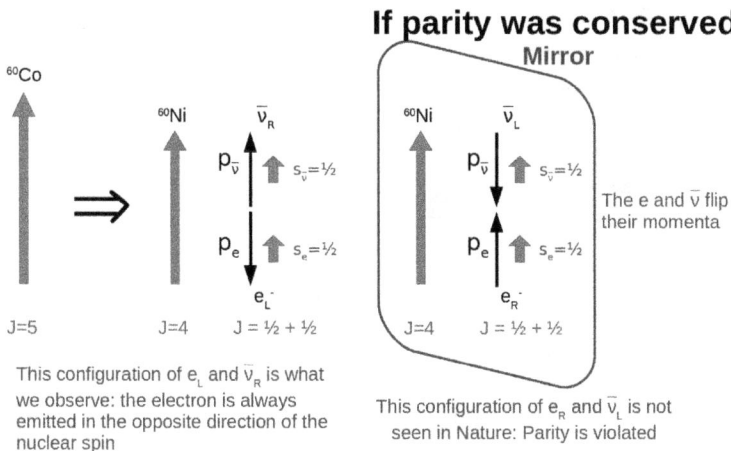

If parity was conserved

This configuration of e_L and $\bar{\nu}_R$ is what we observe: the electron is always emitted in the opposite direction of the nuclear spin

This configuration of e_R and $\bar{\nu}_L$ is not seen in Nature: Parity is violated

What Wu found was that more electrons were emitted in the decay opposite the direction of nuclear spin. Wu estimated that at least 70% of the electrons were emitted in the opposite direction of nuclear spin. Further experiments have observed, in fact, that parity is violated maximally: 100% of the electrons produced in the decay are left-handed, their momentum direction is always opposite to their spin, and no right-handed electrons can ever emerge from this decay.

The beta decay that is taking place is as follows:

$$d \rightarrow u + e_L^- + \bar{\nu}_R, \quad \text{not this one}: \quad d \not\rightarrow u + e_R^- + \bar{\nu}_L.$$

What this tells us is that the weak interaction involves only the left-handed part of quarks and leptons (or the right-handed part of antiquarks and antileptons). Simply put, right-handed particles do not feel the weak interaction. Down quarks have a mass, and they therefore oscillate between the left- and right-handed parts, and decay while in their left-handed "incarnation". Without a mass, the right-handed part would be stable by itself.

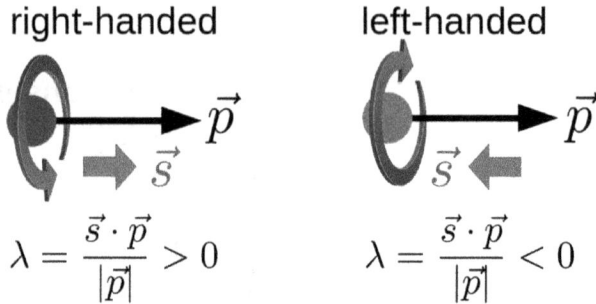

Figure 2.3: The helicity λ of a particle is positive ("right-handed") if the direction of its spin is the same as the direction of its motion. It is negative ("left-handed") if the directions of spin and motion are opposite.

Consider a particle with momentum $\vec{p}$ (a true vector) and spin $\vec{s}$ (an axial vector, since spin is a form of an angular momentum). The projection of the spin on the particle's momentum is referred to as helicity $\lambda = (\vec{s} \cdot \vec{p})/|\vec{p}|$, see Fig. 2.3. Since helicity is defined through a product of a vector and an axial vector, it is an axial.

The current associated with the motion of a particle with positive helicity is called right, and when λ is negative, the current is called left.

In the ultrarelativistic limit (in other words for massless particles), helicity is approximated by another quantum number — chirality (more details in Section A.2.5). If a particle has mass, it is possible to choose a frame of reference such that the particle would appear to be moving in the opposite direction, and helicity would then change its sign. This is not possible for chirality. It is chirality that defines the structure of the weak currents. Figure 2.4 illustrates some examples of chiral and non-chiral objects.

Mathematically, the left current is obtained by acting with a left chirality operator on the Dirac current (see Section A.2. for an introduction to these concepts):

$$\bar{\nu}\gamma_\mu P_L e^- = \bar{\nu}\gamma_\mu \frac{1 - \gamma_5}{2} e^- = \frac{1}{2}\left(\bar{\nu}\gamma_\mu e^- - \bar{\nu}\gamma_\mu \gamma_5 e^-\right) = V - A.$$

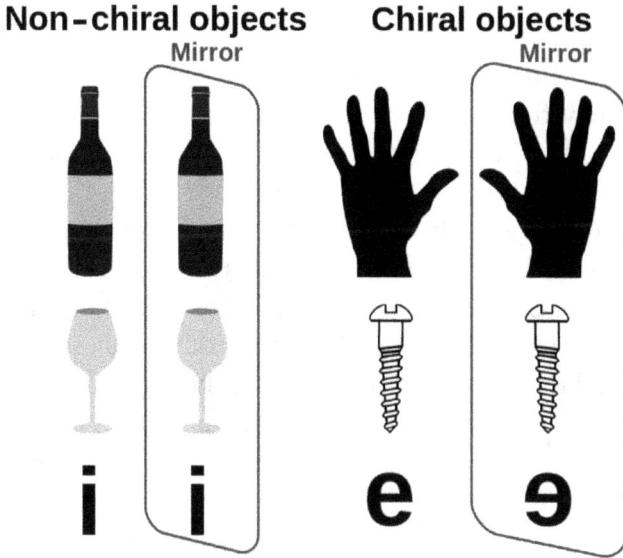

Non-chiral objects   **Chiral objects**

Figure 2.4: If an object and its mirror image can be superimposed and return the same image, the object is said to be "non-chiral". Objects that cannot be superimposed with their mirror image are called "chiral", like our left hand: no matter how many rotations or translations we make to it, we cannot superimpose it with the right hand. Other chiral objects are screws (a right-handed helical thread) or some letters.

Similarly, the right current is obtained by

$$e^+\gamma_\mu P_R \nu = e^+\gamma_\mu \frac{1+\gamma_5}{2}\nu = \frac{1}{2}\left(e^+\gamma_\mu\nu + e^+\gamma_\mu\gamma_5\nu\right) = V + A,$$

where ν/e are Dirac spinors of a neutrino and electron and $\gamma_5 = i\gamma_0\gamma_1\gamma_2\gamma_3$. The first term in these expressions transforms like a vector, while the second one transforms as an axial vector. Based on experimental observations, only left-handed (in terms of chirality) particles and right-handed antiparticles participate in weak charged currents.

2.2   Why Weak Interactions are Weak?

The muon lifetime is similar to that of a charged pion. Muons are found to decay to an electron and two neutrinos. Originally, to explain such a decay, Fermi introduced contact interactions described by a four-fermion vertex (see Fig. 2.5 left). The coupling at such a vertex is $G_F/\sqrt{2}$ with G_F being Fermi constant and a historic factor of $1/\sqrt{2}$. This coupling is small (1.166×10^{-5} GeV^{-2}) and has dimensions of inverse energy squared. It was immediately realized that an introduction of four-fermion vertices is a necessary evil since at large energies they lead to horrible divergences. The strength of that vertex scales as $G_F E^2$, where E is the energy scale for the given process. This dependence poses no problem at low energies, $E \approx m_\mu$ as in the muon decay, but explodes for high energies. So, it was generally agreed upon that the four-fermion vertex is just a low energy approximation and a "better behaved" theory would ultimately replace it (see Box 3.3). If we assume that there is a mediator for decay, called W boson (see Fig. 2.5 right), it could solve the problem with divergences. It would also explain the "weakness" of the weak force, if we assume that it has a large mass. The propagator of a massive boson has a form of

$$\frac{-i\left(g_{\mu\nu} - p_\mu p_\nu/M^2\right)}{p^2 - M^2}, \tag{2.1}$$

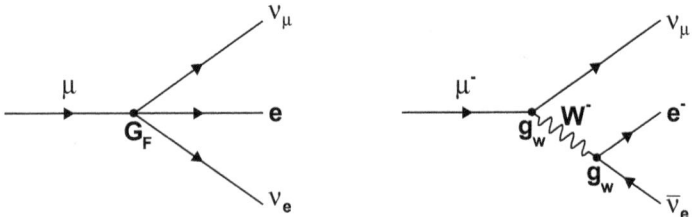

Figure 2.5: Feynman diagrams describing a muon decay to an electron and two neutrinos. Left: four-fermion contact interaction introduced by Fermi. Right: same process mediated by a heavy mediator: the W boson.

where p is the four-momentum of the mediator, M is its mass and the definition of covariant tensor $g_{\mu\nu}$ can be found in Section A.1.1.

When the momentum transfer is small compared to W boson mass ($p \ll M$), the propagator can be replaced with a momentum independent factor $ig_{\mu\nu}/M^2$. If we assume the coupling of the weak interaction g_W to be at the order of electric charge e, we can estimate the expected mass of the mediator:

$$\frac{g_W^2}{M^2} \sim \frac{e^2}{M^2} = \frac{4\pi\alpha}{M^2} \sim \frac{4G_F}{\sqrt{2}} \quad \rightarrow \quad M \sim 25 \text{ GeV}. \qquad (2.2)$$

Here, we use the fine structure constant $\alpha = \frac{e^2}{4\pi} = \frac{1}{137}$. As we will see later this estimate is about a factor of 3 off the true mass of the W boson, yet it correctly sets the scale of weak interactions. Indeed, the momentum transfer at the order of muon mass (105 MeV) is much smaller than the estimated mass of the mediator and the propagator would appear a constant. What we just described is in essence the electroweak symmetry — similar couplings for electromagnetic and weak interactions — and its breaking — the weak interactions are mediated by a massive carrier, while the photon — the mediator of the electromagnetic interaction — is massless.

2.3 The Recyclable SU(2) Group

"Reuse and recycle" is the motto in theoretical physics. Originally, W. Pauli suggested to use the SU(2) group to describe the symmetry between *up* and *down* spin (s) orientation of an electron or any other particle with $s = \frac{1}{2}$. When the neutron was discovered, it was noted that its mass is very close to that of a proton and that both behave very similarly in nuclear scattering reactions. By analogy with spin, the corresponding quantum number was called *isospin*, with the proton being a state with an *up* projection of isospin and neutron being the *down* projection. Isospin $\vec{I}$ is a dimensionless quantity associated with the fact that the strong interaction is ultimately independent of electric charge (cfr. Table 1.1). Any two members of the proton–neutron isospin doublet — proton–proton, proton–neutron, neutron–neutron — experience the same strong interaction. The SU(2) group was thus "recycled" to describe the symmetry properties under the isospin rotation for the strong interaction.

So, when it was observed that electrons and neutrinos behave similarly in the weak interactions, yet another quantum number, weak isospin, $\vec{T}$, was introduced to describe this symmetry. A well-understood SU(2) group was ready to supply predictions for weak reactions. Please note that it is not a vector in the conventional three-dimensional space. It "lives" in its own space of the weak isospin coordinates.

Let us introduce a weak isospinor χ_L such that a neutrino corresponds to a weak isospin spin-up orientation $T_z = +1/2$ and an electron corresponds to weak isospin spin pointing down $T_z = -1/2$

$$\chi_L = \begin{pmatrix} \nu \\ e^- \end{pmatrix}_L, \tag{2.3}$$

where index L denotes left chiral component of the fermion wave function.[1] In the language of SU(2), the weak isospin operator is $\vec{T} = \vec{\tau}/2$, where $\vec{\tau}$ are the Pauli matrices, normally referred to as $\vec{\sigma}$ (Eq. (A.19)). We changed the name to avoid confusion with the Pauli matrices that are applied to "normal" spin. Mathematically though, τ and σ matrices are identical.

Since weak isospin has the same properties as regular spin, we can conclude that the components of $\vec{T}$ have the following commutation relation

$$[T^j, T^k] = i\epsilon_{jkl}T^l, \tag{2.4}$$

where ϵ_{jkl} is a completely antisymmetric tensor.[2]

---

[1]Note that here notation ν or e implies a wave function for the corresponding fermion, which is a four-dimensional Dirac spinor.

[2]As such: $\epsilon_{jkl} = 0$ if any two indices are equal; and $\epsilon_{jkl} = (-1)^n$, where n is the number of transpositions from (123) to (jkl).

Let us construct weak currents by analogy with QED. Recall that the electric current is

$$J_\mu^{\text{EM}} = \bar{\psi}\gamma_\mu \hat{Q}\psi, \tag{2.5}$$

where ψ is the fermion's wave function. The operator $\hat{Q}$ is defined such that:

$$\hat{Q}\psi = Q\psi. \tag{2.6}$$

where the electric charge Q is the integral of the time-like (zeroth) component of the electric current density:

$$Q = \int J_0^{\text{EM}}(x)d^3x. \tag{2.7}$$

Weak isospin is an equivalent of the electric charge as far as weak interactions are concerned. Similarly to Eq. (2.7), the three projections of the weak isospin are the space integrals of the time-like components of the weak current densities:

$$T^i = \int J_0^i(x)d^3x. \tag{2.8}$$

By analogy with Eq. (2.5) we can construct the three-dimensional weak current with a general form of

$$J_\mu^i = \bar{\chi}_L \gamma_\mu T_i \chi_L = \frac{1}{2}\bar{\chi}_L \gamma_\mu \tau_i \chi_L, \tag{2.9}$$

where $i = 1, 2, 3$.

In Pauli formalism, there are the spin up and down operators that are linear combinations of τ_1 and τ_2 matrices, as follows:

$$\begin{aligned} \tau_+ &= \tau_1 + i\tau_2, \\ \tau_- &= \tau_1 - i\tau_2. \end{aligned} \tag{2.10}$$

$W^+(W^-)$ bosons act equivalently to spin-up(down) operators, i.e., the W^+ boson turns an electron into a neutrino ($e_L^- + W^+ \to \nu_L$), and the W^- boson does the opposite ($\nu_L + W^- \to e_L^-$). Charged

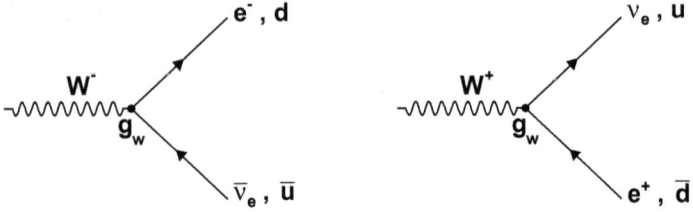

Figure 2.6: Interaction of first-family fermions with $W^\pm$ bosons (charged-current interaction). Identical diagrams can be drawn for the other two families of fermions.

currents of fermions that interact with $W^\pm$ bosons are then defined as follows:

$$J_\mu^+ = \bar{\chi}_L \gamma_\mu \tau_+ \chi_L,$$
$$J_\mu^- = \bar{\chi}_L \gamma_\mu \tau_- \chi_L. \tag{2.11}$$

These are the fermionic currents depicted in Fig. 2.6.

Left-handed particles are members of SU(2) doublets, while right-handed particles, e.g., e_R^- are SU(2) singlets and, as such, do not interact with W bosons.

2.4 Neutral Currents

In Eq. (2.9), we defined three components of the weak current, while only two of them are used to construct the charged currents. Thus, it is logical to assume that there would be a third component of weak current, which would couple to a W^3-boson. Mathematically, the third component of the weak current is described by

$$\bar{\psi}\gamma_\mu \frac{1-\gamma_5}{2} \frac{\tau_3}{2} \psi = V - A, \tag{2.12}$$

where ψ describes the fermion's wave function. Since τ_3 is a diagonal matrix, the flavor of the fermion (or its charge) is not changed in the process, hence we are dealing with a neutral current. For this reason, we will use the term W^0 interchangeably with W^3. The interaction of a W^0 boson with fermions is shown in Fig. 2.7.

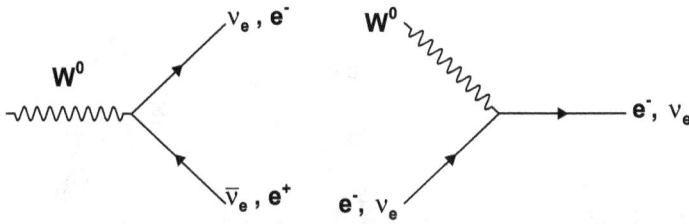

Figure 2.7: Interaction of first-family leptons with W^0 bosons (neutral-current interaction). The left plots illustrate the decay of a W^0 boson producing a lepton–antilepton pair, whereas the right plot shows the absorption of a W^0 boson, changing the momentum and/or spin of the lepton.

The three W bosons would form a triplet in SU(2). This is a state with the weak isospin $T = 1$, with three possible projections $T_z = -1, 0, +1$, corresponding to W^-, W^0 and W^+ bosons, respectively.

Charged currents were well established from the observation of beta decays, so the existence of the $W^\pm$ bosons (named for the *weak force*) was predicted early on. The need for a new neutral particle, and the corresponding neutral current, in order to complete the SU(2) group was by no means obvious. This new particle would be manifested at low energy when neutrinos scatter elastically from matter, but it is not involved in the absorption or emission of electrons and positrons, as the W bosons are. Thus, when neutral W^0 bosons are exchanged, we can only see a change in spin orientation and/or momentum of the particle. This neutral current was predicted independently in the mid-1960s by Sheldon Glashow, Abdus Salam and Steven Weinberg (Fig. 2.8). In 1973, the Gargamelle experiment at CERN observed how a beam of muon neutrinos scattered off electrons: $\nu_\mu + e^- \to \nu_\mu + e^-$, and therefore confirmed the existence of neutral currents (see Box 2.2).

The theory seemed to be coming together. Yet, there was one little problem: more and more evidence seemed to suggest that even though P-parity was not conserved by neutral weak currents, the level of its violation was not the same as for charged currents. In other words, the experimentally observed neutral currents were

Figure 2.8: Sheldon Glashow, Abdus Salam and Steven Weinberg giving seminars at CERN in 1979.

not purely $V - A$, as Eq. (2.12) would suggest, but rather $c_V V - c_A A$, where coefficients c_V and c_A are different from 1. So, the W^0 boson in its pure form cannot be the mediator of the experimentally observed neutral weak interactions. Let us instead refer to it as the Z boson (named for its *zero* electric charge), with the four-dimensional potential of its field being Z_μ.

2.5 Add Photons to Taste

Let us note that the electromagnetic current introduced in Eq. (2.5) is also a neutral current, that is, it does not change the charge of the fermion with which it interacts. We also note that the electromagnetic interactions conserve P-parity, hence the EM current is a pure vector, without an axial component ($c_V = 1$, $c_A = 0$). It can also be presented as the sum of the right and left currents in equal proportion since the cross terms cancel out. For electrons, for example, it can be written as follows:

$$J_\mu^{\text{EM}} = -\bar{e}\gamma_\mu e = -\bar{e}_R \gamma_\mu e_R - \bar{e}_L \gamma_\mu e_L, \qquad (2.13)$$

where the minus sign appears because $\hat{Q}e = -1e$. A few general notes on currents can be found in Section A.5.1.

So, the leap of faith was to state that the experimentally observable neutral bosons — the photon A_μ, mediator of electromagnetic interactions, and Z_μ, mediator of neutral weak interactions — are a mix of W_μ^0 and another neutral boson B_μ, which is a singlet of SU(2). With this construction we can predict the values of coefficients c_V and c_A for Z_μ, and then from the comparison to the experimental data we can evaluate the degree of this mixing.

While the photon couples to electric charge, Q, and W bosons couple to weak isospin, $\vec{T}$, the B_μ couples to a new quantum number called hypercharge, Y. By analogy with the electromagnetic field, we get

$$Y = \int J_0^Y(x)d^3x, \qquad (2.14)$$

where the hypercharge current is defined as

$$J_\mu^Y = \bar{\psi}\gamma_\mu\hat{Y}\psi, \qquad (2.15)$$

and the operator $\hat{Y}$ is defined such that it returns the value of the hypercharge

$$\hat{Y}\psi = Y\psi. \qquad (2.16)$$

We label the strength of the electromagnetic coupling by e, and that of the weak coupling by g. The strength of the hypercharge coupling is $g'/2$. The factor of $1/2$ is historic, but we will keep it for consistency with other literature. We postulate that the hypercharge Y is related to Q and T_3 through the following equation:

$$Q = T_3 + Y/2. \qquad (2.17)$$

Tables 2.1 and 2.2 list the values of the quantum numbers for leptons and quarks. We note that the value of the hypercharge is (and should be) the same for the members of the doublet. From this we conclude that based on Eq. (2.17) the difference in electric charge between the up and down members of the doublet must be $+1$.

From Eq. (2.17) we can derive the hypercharge current:

$$J_\mu^Y = 2J_\mu^{\text{EM}} - 2J_\mu^3. \qquad (2.18)$$

Table 2.1: Quantum numbers for leptons.

Lepton	T	T_3	Q	Y
ν_e, ν_μ, ν_τ	1/2	+1/2	0	−1
e_L^-, μ_L^-, τ_L^-	1/2	−1/2	−1	−1
e_R^-, μ_R^-, τ_R^-	0	0	−1	−2

Table 2.2: Quantum numbers for quarks.

Quark	T	T_3	Q	Y
u_L, c_L, t_L	1/2	+1/2	+2/3	1/3
d_L, s_L, b_L	1/2	−1/2	−1/3	1/3
u_R, c_R, t_R	0	0	+2/3	4/3
d_R, s_R, b_R	0	0	−1/3	−2/3

Plugging Eq. (2.5) and Eq. (2.9) in Eq. (2.18), we get the following expression for the hypercharge current for leptons:

$$J_\mu^Y = -2\bar{e}_R\gamma_\mu e_R - \bar{\chi}_L\gamma_\mu\chi_L. \tag{2.19}$$

2.6 Combining Weak and Electromagnetic Interactions

The electromagnetic interaction is introduced through the addition of the term $-ieJ_\mu^{\text{EM}}A^\mu$ to the Lagrangian, where A^μ is the electromagnetic field four-potential. By analogy, we add terms

$$-igJ_\mu^i W^{\mu i} - i\frac{g'}{2}J_\mu^Y B^\mu, \tag{2.20}$$

to introduce weak isospin and weak hypercharge interactions. The linear combinations of W_μ^1 and W_μ^2 form the charged W bosons

$$W_\mu^+ = \frac{1}{\sqrt{2}}(W_\mu^1 - iW_\mu^2),$$

$$W_\mu^- = \frac{1}{\sqrt{2}}(W_\mu^1 + iW_\mu^2). \tag{2.21}$$

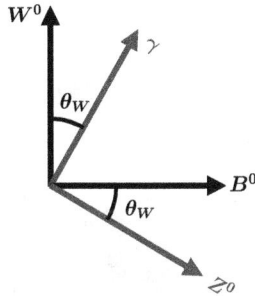

Figure 2.9: The relationship between the neutral vector bosons (mass eigenstates: the photon γ and Z^0 boson) and the neutral gauge fields B_μ^0 and W_μ^0. They are related by the weak mixing angle θ_W, through a simple rotation.

The W_μ^0 and B_μ fields mix with a weak mixing angle θ_W, which at this point is a free parameter of the model, to produce physically observable photons

$$A_\mu = B_\mu \cos\theta_W + W_\mu^0 \sin\theta_W, \tag{2.22}$$

and Z bosons

$$Z_\mu = -B_\mu \sin\theta_W + W_\mu^0 \cos\theta_W. \tag{2.23}$$

This is illustrated in Fig. 2.9.

Solving Eqs. (2.22) and (2.23) for W_μ^0:

$$W_\mu^0 = A_\mu \sin\theta_W + Z_\mu \cos\theta_W, \tag{2.24}$$

and B_μ:

$$B_\mu = A_\mu \cos\theta_W - Z_\mu \sin\theta_W, \tag{2.25}$$

and plugging it in Eq. (2.20), we get the following expression for the neutral currents:

$$-i\left[g\sin\theta_W J_\mu^3 + \frac{g'}{2}\cos\theta_W J_\mu^Y\right] A^\mu$$

$$-i\left[g\cos\theta_W J_\mu^3 - \frac{g'}{2}\sin\theta_W J_\mu^Y\right] Z^\mu. \tag{2.26}$$

The bracket in front of A_μ has to represent the electric current:

$$e J^{\text{EM}} = g \sin \theta_W J^3 + \frac{g'}{2} \cos \theta_W J^Y. \tag{2.27}$$

Multiplying Eq. (2.18) by e, we get

$$e J^{\text{EM}} = e J^3 + \frac{e}{2} J^Y. \tag{2.28}$$

And comparing these two equations we deduce the following relation between coupling constants e, g and g':

$$g \sin \theta_W = e = g' \cos \theta_W. \tag{2.29}$$

Remember that the g is related to the experimentally measurable Fermi constant G_F and the so far unconstrained W-boson mass, m_W (Eq. (2.2)). If we find a way to measure the mixing angle θ_W, the left-hand side of Eq. (2.29) can be used to predict the value of m_W. The right-hand side then constrains the value of coupling g' as follows:

$$g' = g \tan \theta_W. \tag{2.30}$$

Let us now evaluate the coupling strength g_Z and the expression for the current $(J^Z)^\mu$ that interacts with the Z boson using the bracket in front of Z_μ term in Eq. (2.26):

$$g_Z J^Z = g \cos \theta_W J^3 - \frac{g'}{2} \sin \theta_W J^Y = \frac{g}{\cos \theta_W} (J^3 - \sin^2 \theta_W J^{\text{EM}}). \tag{2.31}$$

From this we get the expression for the following coupling:

$$g_Z = \frac{g}{\cos \theta_W}, \tag{2.32}$$

and the current

$$J^Z = J^3 - \sin^2\theta_W J^{\text{EM}}. \tag{2.33}$$

Now, the neutral current is expressed in terms of the neutral component of the $V - A$ current that couples to W bosons, and V current that couples to photons. Using this information we can calculate the experimentally observable coefficients c_V and c_A:

$$J^Z_\mu = \bar\psi\gamma_\mu\left(\frac{1-\gamma_5}{2}T^3 - \sin^2\theta_W Q\right)\psi = \frac{1}{2}\bar\psi\gamma_\mu(c_V - c_A\gamma_5)\psi. \tag{2.34}$$

From this:

$$c_V = T^3 - 2\sin^2\theta_W Q, \tag{2.35}$$

and

$$c_A = T^3. \tag{2.36}$$

To summarize the predictive power of the theory, c_V and c_A of the weak neutral current are evaluated based on the observed asymmetries in nuclear and neutrino scattering experiments. This constrains the mixing angle θ_W. Knowing the Fermi constant and θ_W allows to predict the value of the W-boson mass. The discovery of the W boson with the predicted value of its mass was a stunning triumph of the model of electroweak interactions (see Box 2.2). The present most precise value of $\sin^2\theta_W$ is 0.23121(4), and $m_W = 80.379(12)$ GeV.

Box 2.2 Discovery of the W and Z Bosons (1983)

The first experimental evidence for a unified description of the weak and electromagnetic interactions was obtained in 1973 with the observation of neutral current interactions by the

(Continued)

(*Continued*)

Gargamelle bubble chamber. Neutrinos were imparting momentum to electrons, which could only be explained by the exchange of a virtual, electrically neutral, massive particle:

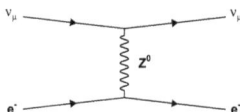

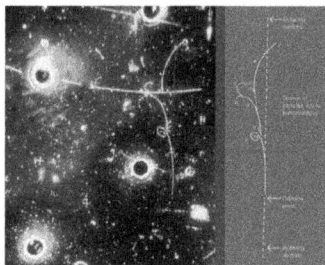

This result provided a determination of the weak mixing angle θ_W, albeit with large uncertainty, and this, in turn, could be used to estimate the mass of the W boson from 60 to 80 GeV and the Z boson mass from 75 to 92 GeV. No accelerator at the time had enough energy to produce such massive particles.

Since LEP construction was still far in the future, it was proposed in 1976 that either CERN or Fermilab could repurpose one of their proton accelerators to a proton–antiproton collider, as a quick and relatively cheap opportunity to reach the energy needed to produce W and Z bosons. CERN approved this concept in 1978 and by July 1981 the Super Proton Synchrotron (SPS) was producing $p\bar{p}$ collisions at $\sqrt{s} = 540$ GeV. The main production process is quark–antiquark annihilation: $u\bar{d} \rightarrow W^+$, $d\bar{u} \rightarrow W^-$, and $u\bar{u} \rightarrow Z$, $d\bar{d} \rightarrow Z$. Approximately half of the energy of a proton is carried by the three valence quarks (uud) and the other half is carried by gluons. Hence, when colliding two quarks, each with $\approx E_{\text{beam}}/6$, one needs a center-of-mass energy at least six times bigger

(*Continued*)

than the W and Z boson masses, or 500–600 GeV. Two detectors, UA1 and UA2, collected enough data by the end of 1982 to observe $W \to e\nu$ decays. And after the spring of 1983, $Z \to e^+e^-$ and $Z \to \mu^+\mu^-$ were also observed. In 1984, Carlo Rubbia and Simon Van der Meer received the Nobel prize "for their decisive contributions to the large project, which led to the discovery of the field particles W and Z, communicators of weak interaction."

These two plots show the accumulated data: on the left is the transverse mass distribution of the W bosons detected in the electron decay mode, corrected for acceptance and resolution. The shaded area shows the background contribution from $W \to \tau\nu$ decays and QCD fluctuations. On the right, the invariant mass distribution for e^+e^- pairs in the Z analysis, where the peak at 90 GeV is clearly separated from the falling QCD background distribution.

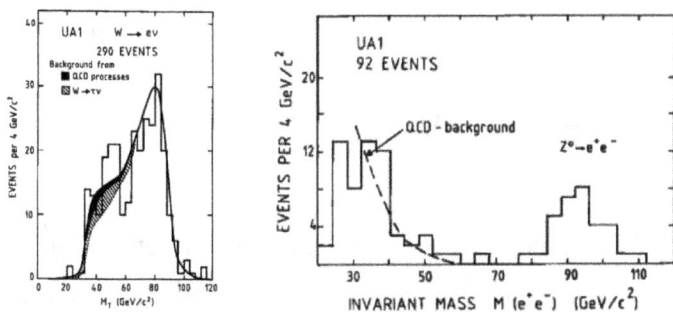

2.7 Feynman Rules

The Feynman diagram for the neutral current interaction is shown in Fig. 2.10. Here the Z boson leg is described by Z_μ. The fermionic

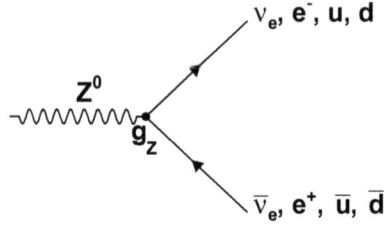

Figure 2.10: Interaction of first-family fermions with the Z boson.

current is described by

$$J_\mu^Z = \bar{\psi}\gamma_\mu\left(\frac{1}{2}(1-\gamma_5)T^3 - \sin^2\theta_W Q\right)\psi = \frac{1}{2}\bar{\psi}\gamma_\mu(c_V - c_A\gamma_5)\psi,$$

(2.37)

and the coupling strength at the vertex is $g_Z = g/\cos\theta_W$.

So, the Feynman rules for weak current interactions are the following:

- each fermion outgoing leg is assigned ψ, each incoming one $\bar{\psi}$;

- each gauge boson is assigned W_μ or Z_μ;

- each vertex carries a factor $-ig_V\gamma_\mu\frac{c_V^f - c_A^f\gamma_5}{2}$, where the coupling g_V is equal to $\frac{g}{\sqrt{2}}$ for charged currents[3] and $\frac{g}{\cos\theta_W}$ for neutral currents, and coefficients c_V^f and c_A^f are equal to 1 for charged currents, while for neutral currents they are given in Table 2.3.

2.8 Properties of W and Z Bosons

The combined theory of weak and electromagnetic interactions asserts that these interactions are mediated by massive W and Z

---

[3]A factor of $\frac{1}{\sqrt{2}}$ arises from the definition of charged W bosons in Eq. (2.21).

Table 2.3: Vector and axial couplings of fermions.

Fermion	c_V^f	c_A^f
$\nu_e,\ \nu_\mu,\ \nu_\tau$	1/2	1/2
$e,\ \mu,\ \tau$	−0.055	−1/2
$u,\ c,\ t$	0.204	1/2
$d,\ s,\ b$	−0.352	−1/2

Table 2.4: W boson decays. The mass of the fermions is neglected.

Decay channel	Color factor	Branching ratio
$e\bar{\nu}_e$	1	1/9
$\mu\bar{\nu}_\mu$	1	1/9
$\tau\bar{\nu}_\tau$	1	1/9
$d\bar{u}$	3	1/3
$s\bar{c}$	3	1/3

bosons and a massless photon. We shall address the issue of generation of boson masses a bit later. For now, let us understand the properties of these particles. The mass of the W boson is constrained by the value of the mixing angle, measured from the asymmetry in neutrino scattering experiments, and the value of Fermi constant, the most precise value of which is provided by the muon lifetime.

The W boson interacts with SU(2) doublets. This means it can decay for example to an electron and an electron antineutrino. All kinematically allowed decays of W bosons are listed in Table 2.4 and shown in Fig. 2.11.

Each quark pair comes in three color combinations (red–antired, blue–antiblue, green–antigreen). The color of quark must match the color of antiquark because W boson is a color neutral state. Thus, there are a total of nine options. W bosons couple "democratically" to all fermion pairs, so in the approximation of zero

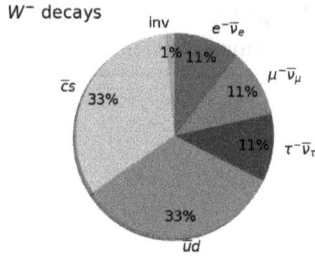

Figure 2.11: Branching ratios of the W boson. Around 1% decays to particles with momentum less than 200 MeV, which are invisible for practical purposes.

fermionic masses these decay options are equally probable. Hence, we can estimate the branching fraction for each decay to be $1/9$ for each pair of leptons and $1/3$ for each pair of quarks.

The situation with Z bosons is a bit more complex since fermions have different values of c_V and c_A. Let us calculate the partial decay width of a generic weak V boson with mass m_V, where V could be W or Z, to a pair of fermions $f\bar{f}'$ with negligible masses. Let p, λ and ϵ_μ be the four-momentum, the helicity and the polarization vector of the V boson, respectively. And let k and s be the momentum and helicity of the fermion f (and k', s' those of fermion f'). Following the Feynman rules, the matrix element for this decay is as follows:

$$i\mathcal{M} = ig_V \bar{f}'\gamma_\mu \frac{c_V - c_A\gamma_5}{2} f\epsilon_\mu \,. \tag{2.38}$$

Averaging over the initial helicity states and summing over the final ones, we get for the matrix element squared the following:

$$\overline{|\mathcal{M}|^2} = \frac{g_V^2}{3} \sum_{\lambda,s,s'} \left(\bar{f}'\gamma_\mu \frac{c_V - c_A\gamma_5}{2} f \right) \epsilon_\mu \epsilon_\nu^* \left(\bar{f}'\gamma_\mu \frac{c_V - c_A\gamma_5}{2} f \right)^* . \tag{2.39}$$

The factor of $1/3$ arises from averaging over three possible helicity states of a spin-1 V boson. We use the following identities for further calculations (see Sections A.2.2 and A.3.2):

$$\sum_{s,s'} (\bar{f}'\gamma_\mu(c_V - c_A\gamma_5)f)(\bar{f}'\gamma_\mu(c_V - c_A\gamma_5)f)^*$$
$$= k_\sigma k'_\tau \text{Tr}[\gamma_\mu\gamma_\sigma\gamma_\nu\gamma_\tau](c_V^2 + c_A^2), \tag{2.40}$$

$$\sum_\lambda \epsilon_\mu \epsilon_\nu^* = g_{\mu\nu} - \frac{p_\mu p_\nu}{m_V^2}. \tag{2.41}$$

The cross terms with γ_5 cancel out because they lead to traces of odd number of γ-matrices that are equal to zero. Thus,

$$\overline{|\mathcal{M}|^2} = \frac{g_V^2}{3}(c_V^2 + c_A^2)\left(g_{\mu\nu} - \frac{p_\mu p_\nu}{m_V^2}\right)(k'_\mu k_\nu - (k'k)g_{\mu\nu} + k_\mu k'_\nu). \tag{2.42}$$

The cross terms containing $p_\mu p_\nu$ produce zero, thus the surviving part of the matrix element squared is as follows:

$$\overline{|\mathcal{M}|^2} = \frac{2g_V^2}{3}(c_V^2 + c_A^2)(k'k). \tag{2.43}$$

Let us now consider the center-of-mass frame of reference. Because of the conservation of energy and momentum, the decay products have an energy equal to half the mass of the parent particle, and the momenta of equal value and opposite direction. Thus,

$$k'k = \frac{m_V^2}{2}. \tag{2.44}$$

So,

$$\overline{|\mathcal{M}|^2} = \frac{g_V^2}{3}(c_V^2 + c_A^2)m_V^2. \tag{2.45}$$

The decay width of a massive particle is given by (see Section A.1.4)

$$\Gamma = \int \frac{1}{2E_V}\overline{|\mathcal{M}|^2}dQ, \tag{2.46}$$

where $E_V = m_V$ is the energy of the parent particle, and

$$dQ = (2\pi)^4 \delta^4 \left(\sum p_i - \sum p_f \right) \frac{d^3k}{(2\pi)^3 2E_f} \frac{d^3k'}{(2\pi)^3 2E_{f'}}, \qquad (2.47)$$

is the Lorentz invariant phase space (LIPS). $E_{f'} = E_f = m_V/2$ are the energies of the final state fermions. The integral over phase space results in a factor of $1/8\pi$, giving the partial width of

$$\Gamma = \frac{g_V^2}{48\pi} (c_V^2 + c_A^2) m_V. \qquad (2.48)$$

To calculate the total decay width, we need to sum over all possible options as follows:

$$\Gamma_{\text{total}} = \sum \Gamma_{f\bar{f'}}. \qquad (2.49)$$

Then the branching ratio for each individual decay is defined as

$$\text{Br}(f\bar{f'}) = \frac{\Gamma_{f\bar{f'}}}{\Gamma_{\text{total}}}. \qquad (2.50)$$

The decay options for W bosons are listed in Table 2.4, while for Z bosons they are listed in Table 2.5 and shown graphically in Fig. 2.12. These results were verified with great precision by LEP experiments using the data collected at the Z resonance. Of particular importance was the measurement of the Z boson invisible width, which constrained the number of neutrino species and limited potential contribution from other hard-to-detect particles, such as dark matter candidates (see Box 2.3). Based on this result we can state that if there is a fundamental particle responsible for dark

Table 2.5: Z boson decays. $\ell = e, \mu, \tau$; $q = u, d, c, s, b$. The branching ratios are calculated at leading order and neglecting fermion masses. The value of $\sin^2 \theta_W$ is 0.2315.

Decay channel	Color factor	Branching ratio
$\ell^+\ell^-$	1	0.094
$\nu_\ell \bar{\nu}_\ell$	1	0.187
$q\bar{q}$	3	0.718

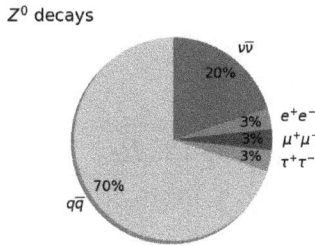

Figure 2.12: Branching ratios of the Z boson.

matter, it must be either heavier than half the Z boson mass, or not couple to Z bosons.

Box 2.3 Z Boson Width

The standard model predicts the partial decay widths of the Z boson to each of the charged lepton species Γ_ℓ, neutrinos Γ_ν and quarks Γ_{had}. One of the first results from the LEP and SLC colliders was the measurement of the number of neutrino generations, N_ν (and thus presumably the number of fermion generations), using data at the Z pole $e^+e^- \to Z \to f\bar{f}$. Assuming that, out of the undiscovered fermions, only neutrinos are lighter than half the Z boson mass, m_Z, we can express its total width as follows:

$$\Gamma_Z = 3\Gamma_\ell + \Gamma_{\text{had}} + N_\nu \Gamma_\nu. \qquad (2.51)$$

The hadron production cross section at the Z peak ($s = m_Z^2$) described by the Breit–Wigner formula from Eq. (A.18),

$$\sigma_{\text{had}} = \frac{12\pi}{m_Z^2} \frac{\Gamma_\ell \Gamma_{\text{had}}}{\Gamma_Z^2}, \qquad (2.52)$$

depends on the total width Γ_Z and leptonic and hadronic widths. We can obtain the number of neutrinos from the

(Continued)

(*Continued*)

previous two equations:

$$N_\nu = \frac{\Gamma_\ell}{\Gamma_\nu} \left(\sqrt{\frac{12\pi R_\ell}{m_Z^2 \sigma_{\text{had}}}} - R_\ell - 3 \right), \qquad (2.53)$$

where we introduced $R_\ell = \Gamma_{\text{had}}/\Gamma_\ell$. With the ratio of leptonic to neutrino widths known from theoretical calculations, the precision in the determination of N_ν is determined mostly by the precision of σ_{had}. Since Z boson decay to hadrons is the dominant mode ($\sim 70\%$), it took only several weeks at the beginning of data taking to measure the number of neutrino species. On October 1989, the SLD and LEP experiments announced that N_ν is consistent with 3, the online average being 3.12 ± 0.19. The figure shows the data of the combined Z boson line shape measured by the four LEP experiments, and what it would look like for 2, 3 or 4 neutrino species. The error bars are enlarged by a factor of 10.

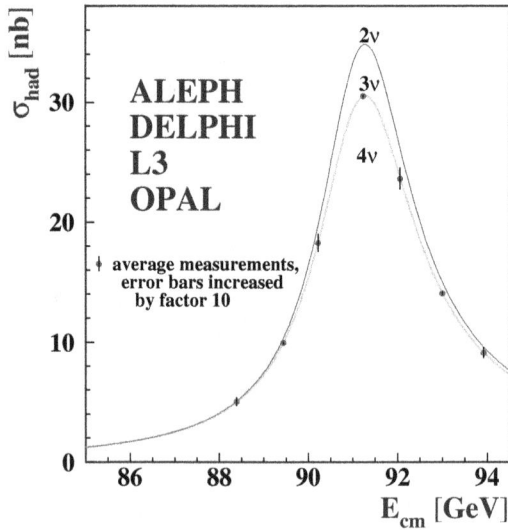

(*Continued*)

The precision of the final result $N_\nu = 2.9841 \pm 0.0083$ is unlikely to be surpassed in the foreseeable future. This result is in remarkable agreement with those obtained from the Big Bang nucleosynthesis.

We note that the described measurement implied that the ratio of the Z boson decay to neutrinos to that to leptons is well described by the standard model. If we want to know if the Z boson decays to other particles that are not registered by conventional experiments and are not a part of the standard model, this assumption needs to be relaxed. It is hard to observe events where there are no observable products. After all, how do we know that a Z boson was produced at all? For this reason, the direct measurement of the Z boson "invisible" width used events where initial state particles irradiate a photon. These studies also did not show any deviations with the standard model, suggesting that any undiscovered particles must be either heavier than half Z boson mass, or do not couple to Z boson, like would right-handed neutrinos should they exist.

2.9 Asymmetries in e^+e^- Annihilation

In electron–positron colliders, the annihilation to a pair of fermions can be mediated by a photon or by a Z boson. Since the initial and final states are identical, the two channels interfere. While the photon-mediated process preserves P parity, the Z-mediated one does not, leading to a beautiful effect of charge asymmetry in the production of fermion pairs in e^+e^- colliders. This effect is sensitive to the value of the weak angle and can be used for its precise measurement.

The matrix element describing $e^+e^- \to \gamma \to f\bar{f}$ is

$$\mathcal{M}_\gamma = -Q_f e^2 (\bar{f}\gamma^\mu f)\frac{g_{\mu\nu}}{k^2}(\bar{e}\gamma^\nu e). \tag{2.54}$$

where k is the four-momentum carried by the mediator. In the center of the mass system $k^2 = s = (2E_{\text{beam}})^2$, where E_{beam} is the beam energy. The matrix element describing $e^+e^- \to Z \to f\bar{f}$ is

$$\mathcal{M}_Z = -\frac{g^2}{4\cos^2\theta_W}[\bar{f}\gamma^\mu(c_V^f - c_A^f\gamma^5)f]\frac{g_{\mu\nu} - k_\mu k_\nu/m_Z^2}{k^2 - m_Z^2}$$
$$\times [\bar{e}\gamma^\nu(c_V^e - c_A^e\gamma^5)e]. \tag{2.55}$$

Recalling that γ^5 is used to construct left and right helicity operators (see Section A.2.2), we can rewrite Eq. (2.55) as[4]

$$\mathcal{M}_Z = -\frac{g^2}{4\cos^2\theta_W(s - m_Z^2)}[c_R^f\bar{f}_R\gamma^\mu f_R + c_L^f\bar{f}_L\gamma^\mu f_L]$$
$$\times [c_R^e\bar{e}_R\gamma_\mu e_R + c_L^e\bar{e}_L\gamma_\mu e_L]. \tag{2.56}$$

where $c_R = c_V - c_A$ and $c_L = c_V + c_A$. Using QED calculation for different helicity amplitudes, we get

$$\frac{d\sigma}{d\Omega}(e_R^+e_L^- \to \bar{f}_R f_L) = \frac{Q_f^2\alpha^2}{4s}(1 + \cos\theta)^2|1 + rc_L^f c_L^e|^2, \tag{2.57}$$

and

$$\frac{d\sigma}{d\Omega}(e_R^+e_L^- \to \bar{f}_L f_R) = \frac{Q_f^2\alpha^2}{4s}(1 - \cos\theta)^2|1 + rc_R^f c_L^e|^2, \tag{2.58}$$

where

$$r = \frac{g^2}{4\cos^2\theta_W(s - m_Z^2 + im_Z\Gamma_Z)}\frac{s}{Q_f e^2}, \tag{2.59}$$

and θ is the polar angle of the fermion f with respect to the electron's direction. For production near the Z pole, it is necessary to

---

[4]We used that in the approximation of zero mass the Dirac equation reads $\frac{1}{2}k_\mu\bar{e}\gamma^\mu = 0$.

account for the finite Z boson width Γ_Z. If beams can be polarized, it is clear that the value of the weak angle θ_W can be extracted from the fit to the angular dependence of the polarized cross sections from Eqs. (2.57) and (2.58). But even for unpolarized beams the sensitivity to the weak angle is preserved since

$$\frac{d\sigma}{d\Omega}(e^+e^- \to \bar{f}f) = \frac{\alpha^2}{4s}[A_0(1 + \cos^2\theta) + A_1\cos\theta], \qquad (2.60)$$

with coefficients A_0 being

$$A_0 = 1 + \frac{1}{2}Re(r)(c_L^f + c_R^f)(c_L^e + c_R^e) + \frac{1}{4}|r|^2(c_L^{f\,2} + c_R^{f\,2})(c_L^{e\,2} + c_R^{e\,2}), \tag{2.61}$$

and A_1 being

$$A_1 = Re(r)(c_L^f - c_R^f)(c_L^e - c_R^e) + \frac{1}{2}|r|^2(c_L^{f\,2} - c_R^{f\,2})(c_L^{e\,2} - c_R^{e\,2}). \tag{2.62}$$

The important feature of the differential cross section in Eq. (2.60) is that it contains both terms that are symmetric and antisymmetric in $\cos\theta$. Hence, we expect the fermion pairs to exhibit the asymmetry between forward and backward production. The level of this asymmetry depends on the value of r, which is sensitive to the weak angle and the mass of Z boson. Using this qualitatively new feature of weak neutral currents the effect of Z boson exchange was observed by PETRA at the center of mass energy of 34 GeV, or way below the Z resonance (see Fig. 2.13).

Measurements of fermion production asymmetries at the Z resonance were used for precise evaluation of the weak angle by LEP and SLC.

2.10 Building the Electroweak Lagrangian

By analogy with QED we have already introduced the Lagrangian terms that describe the interaction of the fermions with weak

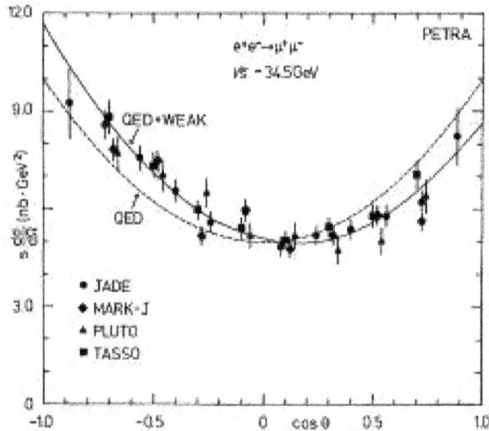

Figure 2.13: Asymmetric distribution in $\cos\theta$ in $e^+e^- \to \mu^+\mu^-$ annihilation that arises due to the exchange of a Z boson, observed by the listed experiments at the PETRA collider in DESY at $\sqrt{s} = 34.5$ GeV.

bosons in Eq. (2.20), but this is not the full story. Let us continue to build on the analogy with QED and formally demand a local gauge invariance in the SU(2) group. The Lagrangian describing a free spin-1/2 particle with mass m has the following form:

$$\mathcal{L}_{\text{free}} = \bar{\psi}(i\gamma_\mu\partial_\mu - m)\psi. \tag{2.63}$$

Applying the Euler–Lagrange formalism (Eq. (A.47)) to this Lagrangian, we get the Dirac equation of motion for a free fermion. It is obvious that a transformation defined as multiplication of the wave function by a factor $e^{iQ\alpha}$, where α is an arbitrary number, does not change this Lagrangian. Thus, invariance under such global gauge (a better name would have been *phase*) transformation is equivalent to the non-observability of the phase of a wave function. If $\alpha(x)$ is a function of a position in space–time, such gauge transformation is called *local*. The requirement that the Lagrangian must be invariant under a local gauge transformation led to the introduction of the electromagnetic field A_μ in QED. The local gauge transformation is defined as follows:

$$\psi(x) \rightarrow e^{i\alpha(x)}\psi(x),$$

$$\partial_\mu \rightarrow D_\mu^{\text{EM}} = \partial_\mu - ieA_\mu, \tag{2.64}$$

$$A_\mu \rightarrow A_\mu + \frac{1}{e}\partial_\mu\alpha.$$

The partial derivative ∂_μ is replaced by D_μ^{EM}, sometimes referred to as the elongated derivative. The full QED Lagrangian invariant under this transformation has a form of

$$\mathcal{L}_{\text{QED}} = \bar{\psi}(i\gamma_\mu\partial^\mu - m)\psi + e\bar{\psi}\gamma_\mu A^\mu\psi - \frac{1}{4}F_{\mu\nu}F^{\mu\nu}, \tag{2.65}$$

where the last term presents the kinetic energy of the electromagnetic field with $F_{\mu\nu}$ defined as

$$F_{\mu\nu} = \partial_\mu A_\nu - \partial_\nu A_\mu. \tag{2.66}$$

Fields that arise as a result of demanding local gauge invariance are referred to as gauge fields. We can check now that the addition of a mass term for the gauge field in the form of $\frac{1}{2}m^2 A_\mu A^\mu$ would break gauge invariance

$$\frac{1}{2}m_\gamma^2 A_\mu A^\mu \rightarrow \frac{1}{2}m_\gamma^2 \left(A_\mu + \frac{1}{e}\partial_\mu\alpha\right)\left(A^\mu + \frac{1}{e}\partial^\mu\alpha\right) \neq \frac{1}{2}m_\gamma^2 A_\mu A^\mu. \tag{2.67}$$

Luckily, photons are massless so we don't need to worry about having such a term, however, when we extend the Lagrangian to include the weak bosons (which we know have large masses and make the weak force relevant for very small distances), these terms would violate gauge invariance, just like the photon mass term here. Fermion masses were introduced in the Lagrangian ad hoc and do not cause a problem with the U(1) gauge invariance, but as we will see, they are not SU(2) invariant.

A carbon copy of this construction with gauge transformation phase $\beta(x)$ can be made for the hypercharge field B_μ, with the conserving hypercharge Y being the equivalent of the conserving electric charge Q. In both cases, the group describing the symmetry is U(1), which is commutative, or Abelian. The situation with three W_μ^j fields ($j = 1, 2, 3$) is slightly different. In SU(2), the phase $\alpha^j(x)$ is a vector in weak isospin space

$$\chi_L \to e^{i\alpha^j(x)T^j}\chi_L. \tag{2.68}$$

Here, $T^j = \frac{\tau^j}{2}$ are the weak isospin operators. The SU(2) symmetry group is non-Abelian, meaning the generators of this group T^j do not commute with each other (see Eq. (2.4)), resulting in terms not canceling in the Lagrangian. To go around this problem, we need to modify the gauge transformation of the gauge field

$$W_\mu^j \to W_\mu^j - \frac{1}{g}\partial_\mu\alpha^j - \epsilon_{jkl}\alpha^k W_\mu^l, \tag{2.69}$$

and the definition of $W_{\mu\nu}$

$$W_{\mu\nu}^i = \partial_\mu W_\nu^i - \partial_\nu W_\mu^i - g\epsilon_{ijk}W_\mu^j W_\nu^k, \tag{2.70}$$

where $i, j, k, l = 1, 2, 3$ are the weak isospin indices. The addition of the last term is not innocent. It leads to the self-interaction of the gauge fields, described by triple and quadruple gauge boson vertices (Fig. 2.14). Self-interacting gauge fields do not show up in QED — there is no scintillating light. Such an effect is a manifestation of the

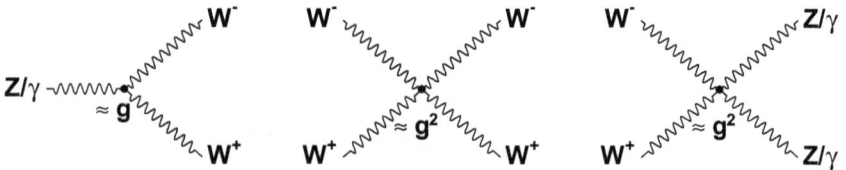

Figure 2.14: Feynman diagrams describing triple (left) and quadruple (center and right) gauge boson vertices.

non-Abelian nature of the group describing the weak interactions and Nature's respect for gauge invariance. Miraculously, reactions resulting from the addition of such terms were indeed observed in experiments, as discussed in Box 2.4.

Box 2.4 The ZWW Vertex is Real

The first observation of the triple gauge boson vertex was made by LEP II experiments in W^+W^- production. This process is not forbidden even in the absence of the triple gauge boson interaction vertex. The W boson is electrically charged and as such couples to photons, thus W^+W^- pair can be produced in s channel in e^+e^- annihilation mediated by a photon. The introduction of the triple gauge boson vertex adds Z boson to the list of mediators in the s channel.

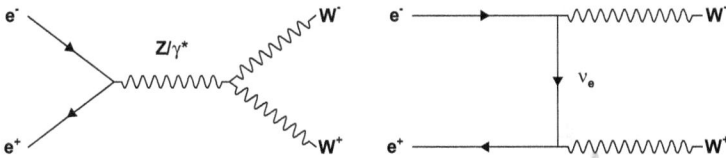

Additionally, this production can happen in the t-channel with the neutrino "running" across (right diagram). Interestingly, the cross sections of both s- and t-channel (see following plot) taken individually grow quadratically with the center-of-mass energy, leading to horrible divergences. However, the two channels interfere destructively leading to just a logarithmic growth with energy suppressed even further by a coefficient proportional to electron mass squared. The observed dependence of the cross section on the energy demonstrated a perfect agreement with the prediction based on the presence of the triple gauge boson vertex.

(Continued)

(*Continued*)

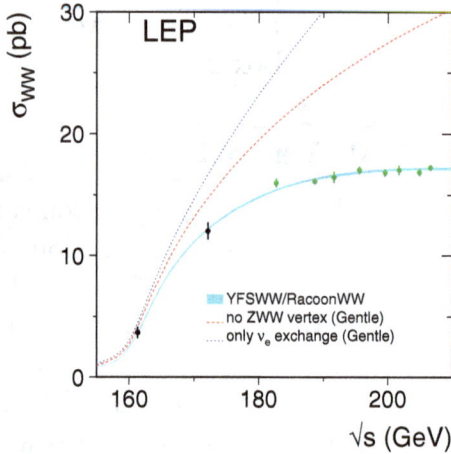

The cross section would be much larger should this contribution not be included.

After the LEP II era, the triple gauge boson interactions were observed by the Tevatron and LHC. The study of center-of-mass energy dependence of these cross sections are on the list of goals of the physics program for High Luminosity upgrade of LHC. The search for triple gauge boson production, which would manifest the quadruple gauge boson vertex, is ongoing.

Thus, the local gauge transformation for electroweak force is defined in the following way:

$$\psi \to e^{iY\beta(x)}\psi,$$

$$\chi_L \to e^{i\alpha^j(x)T^j}\chi_L,$$

$$\partial_\mu \to D^W_\mu = \partial_\mu + igT^jW^j_\mu,$$

$$\partial_\mu \to D^Y_\mu = \partial_\mu + ig'\frac{Y}{2}B_\mu, \tag{2.71}$$

$$W^j_\mu \to W^j_\mu - \frac{1}{g}\partial_\mu\alpha^j - \epsilon_{jkl}\alpha^k W^l_\mu,$$

$$B_\mu \to B_\mu - \frac{1}{g'}\partial_\mu\beta.$$

Table 2.6: Properties of electromagnetic and weak gauge interactions.

Property	EM U(1)	Weak SU(2)$_L$	Weak U(1)$_Y$	Weak Z
Conserved q.n.	Electric charge Q	Weak isospin $\vec{T}$	Hypercharge Y	N/A
Mediator	Photon A_μ	$W_\mu^\pm$ bosons	Hyperfield B_μ	Z_μ boson
Coupling	e	$g/\sqrt{2}$	g'	$g/\cos\theta_W$

To summarize, the Lagrangian describing the electroweak interactions has the following form:

$$\mathcal{L}_{\text{EW}} = \bar{\chi}_L i\gamma_\mu D_\mu^L \chi_L + \bar{\psi}_R i\gamma_\mu D_\mu^R \psi_R - \frac{1}{4}W_{\mu\nu}W_{\mu\nu} - \frac{1}{4}B_{\mu\nu}B_{\mu\nu}\,,$$

$$(2.72)$$

with

$$iD_\mu^L = i\partial_\mu - gT^j W_\mu^j - \frac{g'}{2}Y_L B_\mu,$$

$$iD_\mu^R = i\partial_\mu - \frac{g'}{2}Y_R B_\mu,$$

$$(2.73)$$

where $j = 1, 2, 3$, Y_L is the hypercharge of the left-handed fermionic SU(2) doublet χ_L and Y_R is the hypercharge of the right-handed singlet ψ_R. In Table 2.6, we summarize the properties of the electromagnetic and weak interactions.

We note that there are no mass terms for bosons, such as $\frac{1}{2}m_W^2 W_\mu W^\mu$, because such a term would break even U(1) gauge symmetry. No mass term for fermions of the form: $-\frac{1}{2}m_f^2\bar{\psi}\psi = -\frac{1}{2}m_f^2[\bar{\psi}_R\psi_L + \bar{\psi}_L\psi_R]$ is present either. In SU(2)$_L$, these terms become a problem because ψ_L and ψ_R behave differently under rotations (the first one is left-handed, and as such a member of an isospin doublet, $T = \frac{1}{2}$; whereas the second one is right-handed, hence a singlet $T = 0$), as shown in Eq. (2.71). The product of ψ_R and ψ_L is not SU(2) invariant.

At the same time, we know from experiment that both weak bosons and fermions have non-zero masses. We can either make the Lagrangian not gauge-invariant by adding ad hoc mass terms to account for the measured properties of these particles, or we can leave the theory gauge-invariant but only involving massless particles that we know for a fact have a mass. That is a problem, in fact two separate problems: how to give mass to the fermions and, separately, to the weak bosons, while retaining the symmetries that have guided the theory to such a close phenomenological description of Nature. In the next chapter, we will discuss the solution.

2.11   Gauge Boson Helicity and Mass

Another way to understand the difference between massive and massless gauge bosons is based on helicity considerations. Massless gauge bosons can only have transverse polarization. This fact is well known from classical theory of electromagnetism, where an electromagnetic wave can only be polarized (here, polarization corresponds to the direction of the electric field) in the direction transverse to the direction of motion. Being massless, photons — the quanta of the electromagnetic field — always move with the speed of light. When considering a massive W or Z gauge boson, which moves with a velocity smaller than the speed of light, there is always a possibility to transfer to its frame of reference. There, its angular momentum can have a random direction. If now we transfer to the frame of reference that is moving with a velocity parallel to the angular momentum vector, we observe that the gauge boson has a longitudinal polarization, that is: its spin is parallel or antiparallel to its direction of motion. Hence, while a photon has two degrees of freedom, which correspond to two independent polarizations in the plane transverse to its direction of motion, W and Z bosons possess three degrees of freedom, which also include the longitudinal polarization. The electroweak Lagrangian that we built so far describes only massless gauge bosons, and thus lacks three degrees of freedom needed to describe the physically observed massive W^+, W^- and Z particles. The solution to this problem is described in the next chapter and involves the introduction of the additional degrees of freedom via the introduction of a new field with zero spin.

Suggested Reading for Chapter 2

[1] Chen-Ning Yang and Robert L. Mills. "Conservation of Isotopic Spin and Isotopic Gauge Invariance". *Phys. Rev.* 96 (1954), pp. 191–195.

[2] C. S. Wu et al. "Experimental Test of Parity Conservation in Beta Decay". *Phys. Rev.* 105 (4 Feb. 1957), pp. 1413–1415.

[3] S. L. Glashow. "Partial Symmetries of Weak Interactions". *Nucl. Phys.* 22 (1961), pp. 579–588.

[4] Jeffrey Goldstone, Abdus Salam, and Steven Weinberg. "Broken Symmetries". *Phys. Rev.* 127 (1962), pp. 965–970.

[5] Steven Weinberg. "A Model of Leptons". *Phys. Rev. Lett.* 19 (1967), pp. 1264–1266.

[6] Gerard 't Hooft and M. J. G. Veltman. "Regularization and Renormalization of Gauge Fields". *Nucl. Phys. B* 44 (1972), pp. 189–213.

[7] Gargamelle Neutrino Collaboration. "Observation of Neutrino Like Interactions Without Muon Or Electron in the Gargamelle Neutrino Experiment". *Phys. Lett. B* 46 (1973), pp. 138–140.

[8] UA1 Collaboration. "Studies of Intermediate Vector Boson Production and Decay in UA1 at the CERN Proton — Antiproton Collider". *Z. Phys. C* 44 (1989), pp. 15–61.

[9] ALEPH, DELPHI, L3, OPAL, SLD, LEP Electroweak Working Group, SLD Electroweak Group, SLD Heavy Flavour Group Collaboration. "Precision electroweak measurements on

the Z resonance". *Phys. Rept.* 427 (2006), pp. 257–454. arXiv: hep-ex/0509008.

[10] Steen Hannestad. "Neutrino masses and the number of neutrino species from WMAP and 2dFGRS". *JCAP* 05 (2003), p. 004. arXiv: astro-ph/0303076.

[11] Richard H. Cyburt, Brian D. Fields, and Keith A. Olive. "Primordial nucleosynthesis in light of WMAP". *Phys. Lett. B* 567 (2003), pp. 227–234. arXiv: astro-ph/0302431.

[12] ALEPH, CDF, D0, DELPHI, L3, OPAL, SLD, LEP Electroweak Working Group, Tevatron Electroweak Working Group, SLD Electroweak, Heavy Flavour Groups Collaboration. "Precision Electroweak Measurements and Constraints on the Standard Model" (Dec. 2010). arXiv: 1012.2367 [hep-ex].

[13] ALEPH, DELPHI, L3, OPAL, LEP Electroweak Working Group Collaboration. "Electroweak Measurements in Electron-Positron Collisions at W-Boson-Pair Energies at LEP". *Phys. Rept.* 532 (2013), pp. 119–244. arXiv: 1302.3415 [hep-ex].

Chapter 3

Electroweak Symmetry Breaking

The observation of triple gauge boson vertices proves that SU(2) gauge invariance based on weak isospin is a true symmetry of Nature. However, in imposing this symmetry on the Lagrangian, one is faced with two separate yet equally challenging problems: gauge boson masses and fermion masses.

The solution to both problems comes from the introduction of a scalar complex SU(2) doublet field ϕ with a non-zero vacuum expectation value. The scalar field is named after one of its parents, Peter Higgs. The Lagrangian remains SU(2) invariant, the potential that describes the scalar field is also gauge invariant, but the particular choice of the ground state of this potential is not. A good analogy for this is a pencil standing on its tip. The Lagrangian describing this system is invariant under the rotation by an azimuthal angle. Yet, when the pencil falls, it *spontaneously* chooses a particular ground state, thus breaking the azimuthal symmetry (Fig. 3.1).

3.1 Scalar Field Doublet and Its Quantum Numbers

We introduce a scalar ($s = 0$) complex SU(2) doublet field

$$\phi = \sqrt{\frac{1}{2}} \begin{pmatrix} \phi_1 + i\phi_2 \\ \phi_3 + i\phi_4 \end{pmatrix}, \tag{3.1}$$

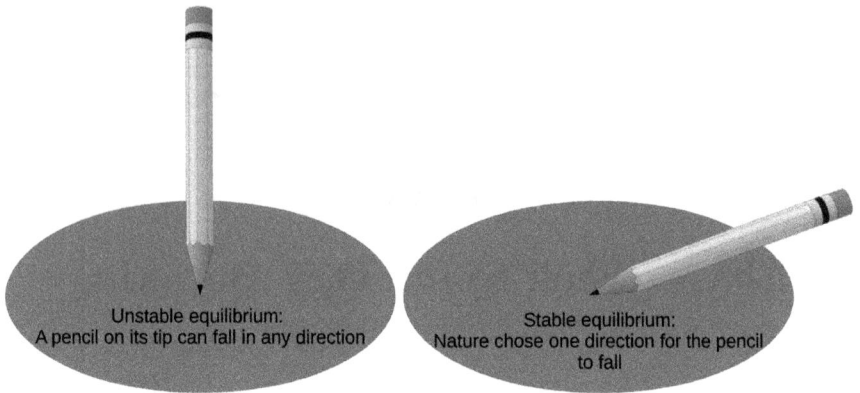

Figure 3.1: Spontaneous symmetry breaking in Nature: a pencil on its tip can fall in any direction, the physics describing this problem has to be symmetric on the azimuthal angle (left). But somehow Nature breaks that symmetry and the pencil does fall in a given direction (right). A similar example is when a dining guest at a circular table is the first to pick the bread on their left or their right: before there was symmetry, after the symmetry is broken by the first guest, everyone else will have to pick in the same direction.

described by four degrees of freedom $(\phi_1, \phi_2, \phi_3, \phi_4)$. As an SU(2) doublet it has the value of weak isospin $T = 1/2$, with $(\phi_1 + i\phi_2)$ corresponding to an *up* projection of weak isospin $(T_z = +1/2)$ and $(\phi_3 + i\phi_4)$ being a *down* projection $(T_z = -1/2)$. We noted previously that based on Eq. (2.17) the difference in electric charge between the *up* and *down* members of the doublet must be $+1$. We choose $(\phi_1 + i\phi_2)$ to have positive electric charge $Q_{up} = +1$ and $(\phi_3 + i\phi_4)$ to be electrically neutral $Q_{down} = 0$. The value of the hypercharge of the scalar field is then $Y = 1$.

We introduce a potential of this field in the following form:

$$V(\phi) = \mu^2 \phi^\dagger \phi + \lambda(\phi^\dagger \phi)^2, \qquad (3.2)$$

where μ^2 and λ are the free parameters of the theory. The fourth power potential was originally introduced by Ginzburg and Landau in 1939 to explain the second-order phase transitions. It was later applied to explain the symmetry breaking in solid state crystals, as well as superconductivity (see Box 3.2).

The terms in the Lagrangian that describe this new scalar field in its free state are

$$\mathcal{L}_h = (\partial_\mu \phi)^\dagger (\partial_\mu \phi) - \mu^2 \phi^\dagger \phi - \lambda (\phi^\dagger \phi)^2. \tag{3.3}$$

The first two terms describe a free particle of mass μ^2, and the λ term describes the four-point self-interaction. Since ϕ is an SU(2) doublet, we expect it to interact with the gauge boson fields. Thus, the Lagrangian including the interaction terms of the Higgs field with the gauge bosons has the form

$$\mathcal{L}_h = \left(iD_\mu^h \phi\right)^\dagger \left(iD_\mu^h \phi\right) - V(\phi),$$
$$iD_\mu^h = i\partial_\mu - gT^j W_\mu^j - g'\frac{Y}{2}B_\mu. \tag{3.4}$$

The form of the elongated derivative D_μ^h is identical to D_μ^L from Eq. (2.72), since it describes the interaction of an SU(2) doublet with the gauge fields, yet the subscript L is meaningless when applied to zero-spin fields.

3.2 The Higgs Potential

The fourth power of the potential in Eq. (3.2) is specifically chosen because it allows for a non-zero vacuum expectation value (vev) v. As shown in Fig. 3.2, depending on the relative sign of μ^2 and λ the potential can have a global minimum at $|\phi| = 0$ (thus having a zero vev), or a local maximum at $|\phi| = 0$ and a global minimum at $|\phi| = \sqrt{\frac{-\mu^2}{2\lambda}} = \frac{v}{\sqrt{2}}$. It is logical to assume that the charged component $(\phi_1 + i\phi_2)$ has zero vev (after all, it would be very uncomfortable to have a positively charged vacuum), while the neutral part of the doublet $(\phi_3 + i\phi_4)$ acquires non-zero vev. Since ϕ is a complex number, the minimum of the potential is a continuum of states described by $\phi = \frac{v}{\sqrt{2}}e^{i\eta}$, where $0 < \eta < 2\pi$ (see Fig. 3.3). Nature spontaneously chooses a particular value of η, when she rolls into the minimum of the potential, similarly to how a pencil placed on its tip chooses a particular angle at which to fall. In other words,

$$V(\phi) = \mu^2\phi^\dagger\phi + \lambda(\phi^\dagger\phi)^2$$

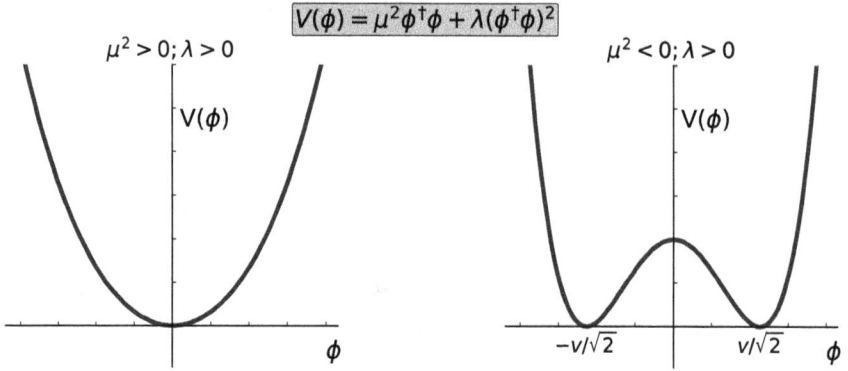

Figure 3.2: The Higgs potential $V(\phi)$ for different signs of the parameter μ^2. The value of the vev is defined as $v^2 = -\mu^2/\lambda$.

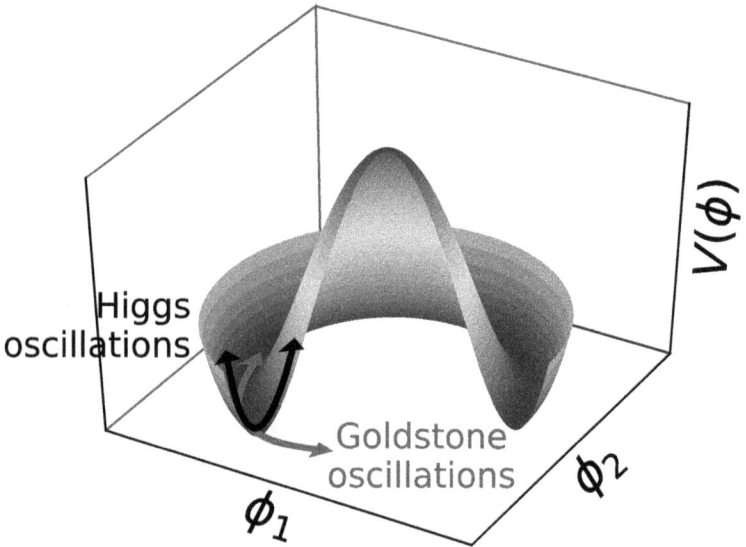

Figure 3.3: The Higgs potential $V(\phi)$ as a function of the real (ϕ_1) and imaginary (ϕ_2) parts of the field ϕ. Once the field falls in the minimum, it can oscillate longitudinally (around the rim) or transversely (up and down the rim).

an SU(2) symmetry is spontaneously broken by the choice of the ground state. We can define the value of $\eta = 0$ to correspond to this choice. Then, the potential reaches its minimum at the following value of the Higgs field:

$$(\phi)_{\min} = \frac{1}{\sqrt{2}} \begin{pmatrix} 0 \\ v \end{pmatrix}. \tag{3.5}$$

Plugging Eq. (3.5) in Eq. (3.4) and using $T^j = \frac{\tau^j}{2}$ and $Y = 1$ we get the Lagrangian of the Higgs and gauge bosons in its ground state

$$\mathcal{L}_h = \frac{1}{8} \left| (2i\partial_\mu - g\tau^j W^j_\mu - g'B_\mu) \begin{pmatrix} 0 \\ v \end{pmatrix} \right|^2 - V(\phi). \tag{3.6}$$

3.3 Generating Gauge Boson Masses

Concentrating on the term describing W and B interactions[1] with the scalar field in Eq. (3.6) and using the Pauli representation of τ^j matrices we get

$$(iD^h_\mu \phi)^\dagger (iD^h_\mu \phi) \rightarrow \frac{1}{8} \left| \begin{pmatrix} gW^3 + g'B & g(W^1 - iW^2) \\ g(W^1 + iW^2) & -gW^3 + g'B \end{pmatrix} \begin{pmatrix} 0 \\ v \end{pmatrix} \right|^2. \tag{3.7}$$

The product of the 2×2 matrix and the spinor results in a spinor. Once squared it is a sum, which has two terms

$$\frac{v^2}{8} \left(g^2 \left[(W^1)^2 + (W^2)^2 \right] + \left[-gW^3 + g'B \right]^2 \right). \tag{3.8}$$

One term involves W^1 and W^2, which can be recast as a mass term for W^+ and W^- bosons

$$\frac{g^2 v^2}{8} \left[(W^1)^2 + (W^2)^2 \right] = \frac{1}{2} m_W^2 \left[(W^+)^2 + (W^-)^2 \right]. \tag{3.9}$$

Since we expect the general mass term for a massive gauge boson V to be of the form $\frac{1}{2} m_V^2 V_\mu^2$, we are able to equate the

---

[1] Here and for the rest of the chapter, we drop the indices μ for gauge bosons for compactness of notation.

coefficient in front of $(W^\pm)^2$ as the physical $W^\pm$ boson mass

$$\frac{g^2 v^2}{4} = m_W^2 \Rightarrow m_{W^+} = m_{W^-} = \frac{gv}{2}, \tag{3.10}$$

which allows us to get a prediction for the value of the vev given m_W

$$v = \frac{2m_W}{g} = 246 \text{ GeV}. \tag{3.11}$$

The second term in Eq. (3.8) can be written as the mass term for the Z boson

$$\frac{(g^2 + g'^2)v^2}{8}(\cos\theta_W W^3 - \sin\theta_W B)^2 = \frac{1}{2}m_Z^2 Z^2, \tag{3.12}$$

where we have equated

$$m_Z = \frac{v}{2}\sqrt{g^2 + g'^2}, \tag{3.13}$$

and introduced the electroweak angle as shorthand for the ratio of the W and Z boson masses, in terms of the couplings

$$\frac{m_W}{m_Z} = \cos\theta_W = \frac{g}{\sqrt{g^2 + g'^2}}. \tag{3.14}$$

This prediction is experimentally verifiable through the observation of a Z gauge boson at the right mass. This indeed happened, when both W and Z bosons were discovered at the SPS collider at CERN in 1983 with masses in spectacular agreement with the theory prediction (see Box 2.2).

We note that the linear combination of W^3 and B fields corresponding to a photon remains massless

$$0 \cdot (\sin\theta_W W^3 + \cos\theta_W B)^2 = \frac{1}{2}m_\gamma^2 A^2. \tag{3.15}$$

So, the problem of introducing gauge boson masses in the Lagrangian describing the weak interactions is solved by introducing a fundamental scalar field that is an SU(2) doublet and a fourth power potential associated with it. Depending on the relative sign between the coefficients of this potential it can develop a minimum at a non-zero value of the field. The field, the potential, and its continuum of minima are SU(2) invariant, yet the spontaneous choice of a particular minimum breaks the SU(2) symmetry, hence this process is referred to as a spontaneous symmetry breaking. In this process, charged W bosons and one particular combination of the neutral W and B fields, corresponding to the Z field, acquire mass via their interaction with the non-zero vev of the scalar field. The orthogonal combination corresponding to the electromagnetic field remains massless. Hence, it is also possible to say that the symmetry between weak and electromagnetic interactions is broken.

A historic perspective on this theoretical construction is presented in Box 3.1.

Box 3.1 A Unified Electroweak Theory (1967)

The phenomena described by electromagnetism and the weak interactions don't appear to have much in common at all. The former holds atoms together, produces light, and on the surface is part of our everyday experience, the latter at the beginning was only familiar due to beta decay in nuclear radioactivity. However, both kinds of interactions act on leptons and hadrons and both appear to be *vector* interactions transmitted by the exchange of particles with spin one and negative parity. In 1961, Sheldon Glashow attempted to describe these two interactions in a unified theory based on the gauge invariance of SU(2)×U(1). This was based on the work by Chen Ning Yang and Robert L. Mills who were extending gauge invariance from Abelian groups like U(1) for electromagnetism, to non-Abelian SU(n) group, in hopes of describing the strong interaction. Remember, in Abelian groups the order in which gauge transformations are applied does not matter, whereas in non-Abelian groups the result depends on the order. This gives the theory a more complicated mathematical structure

(Continued)

(*Continued*)

but also opens up new possibilities. For example, this allowed the charged current, but not the neutral current (electromagnetism), to violate parity.

Of course, the problem remained of how to describe a massless photon as the mediator of electromagnetism (the range of the force is infinite) while needing a massive spin-1 mediator for weak interactions (beta decay is limited to subatomic distances). Glashow's version of the theory could not find an easy way of including massive weak mediators without breaking the gauge invariance that was the main motivation to start with an SU group. This added feature would come from the work on spontaneous symmetry breaking by Brout, Englert, Higgs, Kibble, Guralnik and Hagen. They demonstrated that the vector bosons of a Yang–Mills theory can acquire a mass without spoiling the fundamental gauge symmetry.

Weinberg had been trying to incorporate spontaneous symmetry breaking on an SU(2)×SU(2) theory of strong interactions, but the rho-mesons in the theory remained massless, contrary to reality. Weinberg decided then to describe a model of leptons only. One starts from two left-handed doublet (the ν_{eL} and e_L) and a right-handed singlet (the e_R), which can naturally fit in a SU(2)×U(1)×U(1) for one of the U(1) groups to be identified with a lepton number. But given that lepton number is conserved and no massive vector particle couples to it, it may be excluded from the group. And that leaves the SU(2)×U(1) with four parameters. By spontaneously breaking SU(2)×U(1) to the U(1) of ordinary electromagnetic gauge invariance, three of the four vector gauge bosons acquire a mass: the charged $W^\pm$ and the neutral boson Z^0, leaving the photon massless. In 1967, this was just a possible model, a hypothesis, and it was in fact disregarded until it was seen that it was, in fact, renormalizable (Box 3.3).

(*Continued*)

The best way to prove if such a model was correct, was to find the weak neutral current, which was achieved by the Gargamelle experiment in 1973 (Box 2.2). But the W boson itself would not be discovered until 1983!

3.4 Generating Fermion Masses

We pointed out that not only do gauge boson masses present a problem with the SU(2) symmetry of the Lagrangian, as it would have been the case for the QED, fermion masses, too, cannot be added to the Lagrangian without breaking the SU(2) symmetry. The newly introduced scalar field was charged with solving this problem as well. Since Higgs field is an SU(2) doublet, its scalar product with another SU(2) doublet (take neutrino and electron for example) is an SU(2) singlet. So terms of the form

$$-g_e \left[(\bar{\nu}_e, \bar{e})_L \begin{pmatrix} \phi^+ \\ \phi^0 \end{pmatrix} e_R + \bar{e}_R (\phi^-, \bar{\phi}^0) \begin{pmatrix} \nu_e \\ e \end{pmatrix}_L \right] \tag{3.16}$$

are SU(2) invariant and can be added to the Lagrangian without any problem. Here, we introduced a new free parameter g_e which describes the strength of the coupling between the scalar and the fermionic fields. Once we plug in the value of the scalar field corresponding to the minimum of the potential from Eq. (3.5), such terms will have the form of the mass term for fermions

$$-\frac{g_e v}{\sqrt{2}} (\bar{e}_L e_R + \bar{e}_R e_L) = -m_e \bar{e} e. \tag{3.17}$$

Since only the "lower" component of the Higgs doublet acquires non-zero vev, this mechanism of mass generation works well for the fermions that correspond to the weak isospin *down* orientation. Note that since no right-handed neutrinos ν_R have been discovered, the Higgs mechanism cannot be used to generate neutrino masses.

For upper members of the quark doublets we construct a new scalar doublet field

$$\phi_c = -i\tau_2\phi^* = \begin{pmatrix} -\bar{\phi}_0 \\ \phi^- \end{pmatrix}. \tag{3.18}$$

Once the SU(2) symmetry is broken, field ϕ_c "rolls" into its minimum

$$(\phi_c)_{\min} = \sqrt{\frac{1}{2}} \begin{pmatrix} v \\ 0 \end{pmatrix}. \tag{3.19}$$

Thus, the up-type quark mass term has the form

$$-g_u(\bar{u}, \bar{d})_L \begin{pmatrix} -\bar{\phi}^0 \\ \phi^- \end{pmatrix} u_R \rightarrow -\frac{g_u v}{\sqrt{2}} \bar{u}u \tag{3.20}$$

Note, that constants g_f, referred to as Yukawa couplings, must be individually "tailored" for each fermion, such that the fermion's mass, m_f, is described by

$$m_f = \frac{g_f v}{\sqrt{2}}. \tag{3.21}$$

This is a far less appealing feature of the model. The number of free parameters is not reduced, since the suite of unconstrained fermionic masses is traded for equally unconstrained suite of fermionic couplings with the Higgs field.

3.5 Higgs Boson Mass and Self-Interaction

The second power term in the Higgs potential is equivalent to the mass term of the Higgs field itself

$$\mu^2\phi^\dagger\phi = -\frac{1}{2}m_h^2\phi^\dagger\phi, \tag{3.22}$$

with the Higgs mass being

$$m_h = v\sqrt{2\lambda}. \tag{3.23}$$

While the value v was constrained by the W boson mass, the self-interaction coefficient λ remained unconstrained until the discovery of the Higgs boson, which complicated the search since theory did not predict its mass.

3.6 Excitation Around the Minimum — Higgs the Particle

We introduced a new field ϕ with a non-zero vev, which was extremely helpful in generating masses for gauge bosons. Yet, in quantum mechanics one cannot pin down the field in its minimum, there will always be excitations around the minimum

$$\phi = \frac{1}{\sqrt{2}} \begin{pmatrix} 0 + i\pi_+(x) \\ v + h(x) + i\pi_0(x) \end{pmatrix}, \tag{3.24}$$

where v is a constant and $h(x)$, $\pi_+(x)$ and $\pi_0(x)$ are excitations of the Higgs field around the minimum of its potential illustrated in Fig. 3.3. Excitations "along the rim" of this potential generate the so-called massless Nambu–Goldstone bosons: $\pi_+(x)$ and $\pi_0(x)$. These additional degrees of freedom are exactly what is needed to provide the longitudinal polarizations of the now massive W and Z bosons discussed in Section 2.11. A scalar complex Higgs doublet is described by four degrees of freedom, three of which are absorbed by the longitudinal polarizations of the W and Z bosons, leaving one degree of freedom, the excitation transverse to the rim. This is the physical scalar particle $h(x)$ called the Higgs boson.

Plugging the expansion around the minimum from Eq. (3.24) in the electroweak Lagrangian of Eq. (3.4), generates a variety of interactions between the $h(x)$ field and the gauge bosons, the fermions and itself. The part of the Lagrangian that includes the terms describing the interaction of the h field with W and Z bosons has the form

$$\mathcal{L}_{hWZ} = \frac{g^2}{4}(2hv + h^2)W^+W^- + \frac{g^2 + g'^2}{8}(2hv + h^2)Z^2, \tag{3.25}$$

where the coefficient in front of the W boson term is twice as large compared to the Z boson term because of the definition of W^+ and W^- given in Eq. (2.21). The interaction of the Higgs field with the fermions is described by

$$\mathcal{L}_{hf} = \frac{g_f}{\sqrt{2}} h \bar{f} f. \tag{3.26}$$

Table 3.1: Vertex factors for Higgs boson interactions with gauge bosons, fermions and with itself. Whenever there are n identical bosons, there is an extra factor of $n!$

Interaction	Vertex factor $(-ig_{\mu\nu})$	Diagram
hW^+W^-	$\dfrac{g^2}{2}v$	
hZZ	$2!\dfrac{g^2+g'^2}{4}v$	
$h\bar{f}f$	$\dfrac{g_f}{\sqrt{2}}$	
hhW^+W^-	$2!\dfrac{g^2}{4}$	
$hhZZ$	$2!2!\dfrac{g^2+g'^2}{8}$	
hhh	$3!\lambda v$	
$hhhh$	$4!\dfrac{\lambda}{4}$	

And finally, the interaction of the Higgs with itself in triple and quartic vertices takes the form

$$\mathcal{L}_{hh} = -\frac{\lambda}{4}(4vh^3 + h^4). \tag{3.27}$$

In the language of the Feynman diagrams, each new term adds another vertex. The factors associated with each such vertex are expressed in terms of coupling constants, v and λ (see Table 3.1), or in other words, the strength of the Higgs boson interaction with other particles is predicted within the standard model given the values of v and λ, or alternatively the values of v and μ. As we have seen the value of v is constrained by the value of the W boson mass (or alternatively by $\sin\theta_W$ and G_F), while the second constant was not constrained prior to Higgs boson discovery. Some of these newly introduced interactions of the Higgs boson were used as a tool for its discovery; some were observed later; and some, in particular related to Higgs boson self-interaction, are waiting for their hour.

Box 3.2 Food for Thought

The introduction of a scalar field that is an SU(2) doublet allowed to solve problems in one go — introduce masses for gauge bosons and fermions in a gauge invariant way. We emphasize that even though the scalar field played a central role in both cases, the solutions to these problems are quite different. While in the first case, the non-zero vacuum expectation of the scalar fields was absorbed to generate the degrees of freedom corresponding to the gauge bosons longitudinal polarization, in the second case, a Yukawa-like coupling had to be introduced to generate fermionic masses. The first construction is highly predictive: it provided with a m_Z/m_W prediction (Eq. (3.14)), as well as a connection between the strength of the interaction of the Higgs field and the masses

(Continued)

(*Continued*)

of the gauge bosons (Eqs. (3.10) and (3.13)). The second construction just recasted fermion masses into fermion Yukawa couplings (Eq. (3.21)), without reducing the number of free parameters. It is not clear why the Yukawa coupling would be custom tailored for each fermion. It is not clear why there is such a range of Yukawa couplings, starting with $\sim 10^{-5}$ for the electron and first generation quarks and going all the way to 1 for the top quark.

Even more general questions are why is there a scalar field only for weak interactions, why not for strong interactions; why the generation of fermionic mass is tied to weak interactions; and is there a connection between parity violation and mass generation?

The Higgs field was introduced based on an analogy with the Landau–Ginzburg order parameter from the theory of superconductivity. Now we know that this theory is only effective on large scales. There is a microscopic interpretation of the order parameter via the introduction of electron Cooper pairs. Is there a similar mechanism for the Higgs field. Could it be composite?

The exploration of scalar field(s) has just begun. These and other questions will be answered by future generations of physicists.

3.7 Summary of Theory Predictions

We postulated the existence of a fundamental scalar doublet ϕ ($s = 0$, $P = +1$) and a potential associated with it

$$V(\phi) = \mu^2 \phi^\dagger \phi + \lambda(\phi^\dagger \phi)^2. \tag{3.28}$$

This potential has a non-zero vacuum expectation $v = \sqrt{\frac{-\mu^2}{\lambda}} = 246$ GeV. Gauge bosons and fermions interact with this field and acquire mass as a result. Hence, v sets the scale for masses of all fundamental particles. The W boson mass is

$$m_W = \frac{g}{2}v. \tag{3.29}$$

The Z boson mass is

$$m_Z = \frac{g}{2\cos\theta_W}v. \tag{3.30}$$

The masses of fermions are given by

$$m_f = \frac{g_f}{\sqrt{2}}v. \tag{3.31}$$

Finally, the Higgs boson itself acquires mass via self-interaction

$$m_h = v\sqrt{2\lambda}. \tag{3.32}$$

As we see, each of these masses is proportional to v with the coefficient that describes the strength of the particle's coupling with the scalar field. Experimentally, g, $\cos\theta_W$ and the fermion masses can be measured. Yet, the self-coupling λ was not determined before the Higgs discovery. Hence, all the prediction of Higgs boson properties, e.g., production cross sections, total width, and branching ratios were done as a function of the scalar boson mass.

Box 3.3 The Electroweak Interaction is Renormalizable (1972)

Early on in the development of quantum field theory for electromagnetism (QED), it was realized that problems arose when doing calculations taking into account the radiative self-energies of particles. For example, an electromagnetic field near a particle (say, an electron) can spontaneously generate copious amounts of *virtual particles* (short-lived pairs of particles and antiparticles). Thus, the original "naked" single electron turns out to be surrounded

(Continued)

(*Continued*)

("dressed") by a myriad of other electron–positron pairs popping in and out of the vacuum, like shown here:

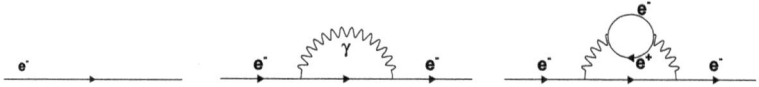

This fact yielded ultraviolet divergent results when doing the detailed calculations on the energy of such an electron. The solution for QED was proposed in the 1940s by Sin-Itiro Tomonaga, Julian Schwinger and Richard P. Feynman (who shared the Nobel prize in 1965 for this contribution). They introduced the idea of *renormalization* for an Abelian gauge theory, by which all infinities disappear if one takes into account that the original electron has, in fact, a new mass and charge (not the ones from the Lagrangian) as derived from including the clouds of virtual photons and electron–positron pairs surrounding it. Basically, one can look at the electron "from a distance", where the original electron appears covered in the virtual pairs. By using this newly derived mass and charge with the renormalization techniques, suddenly all sorts of calculations become possible and are confirmed to agree with experimental measurements!

Broadly speaking, a theory is renormalizable as long as the coupling constants do not have dimensions of negative powers of mass. In fact, only a few simple types of interactions turn out to be renormalizable because any added field or space-time derivative has the effect of reducing the dimensionality of the associated coupling constant. For example, the Fermi theory of weak interactions was not renormalizable, given that G_F has dimension of $[mass]^{-2}$ (see Eq. (2.2)).

When a non-Abelian gauge theory was introduced to unify electromagnetism and the weak interactions (see Box 3.1), the

(*Continued*)

same problem arose: the theory predicted the existence of W and Z bosons, but their calculated properties gave infinities. By 1967, deWitt, Faddeev, Popov and others had proven that non-Abelian gauge theories with unbroken symmetries were renormalizable. The question was whether the introduction of spontaneous symmetry breaking in such a theory would spoil the renormalizability.

In 1971, a Dutch graduate student named Gerardus 't Hooft and his advisor Martinus Veltman (Fig. 3.4) solved this issue by inventing a new gauge, like the "Feynman gauge" in QED (Eq. (2.64), which is used throughout this book). Each choice of gauge fixing leads to different sets of Feynman rules (Section 2.7), which basically encode how each Feynman diagram enters the calculations. Veltman and 't Hooft introduced a gauge in which the Feynman rules can be made to only lead to a finite number of types of ultraviolet divergence.

The Feynman diagram (and its corresponding rules) can now be used to calculate observables in perturbation theory. If they lead to divergent integrals, it's all a matter of finding the appropriate masses and couplings of the theory such that divergent expressions can be absorbed by redefinitions of these parameters.

Figure 3.4: Martinus Veltman in 1973 and Gerardus 't Hooft visiting ATLAS in 2009.

Suggested Reading for Chapter 3

[1] Heather E. Logan. "TASI 2013 lectures on Higgs physics within and beyond the Standard Model" (June 2014). arXiv: 1406.1786 [hep-ph].

Chapter 4

Higgs Boson Production and Decay

In this chapter, we discuss Higgs boson production in lepton and hadron colliders and its possible decays. These predictions are done as a function of one free parameter, the Higgs boson mass, which the theory did not predict a priori.

4.1 Higgs Decays

At tree-level, the Higgs boson, which is an excitation of the scalar field around the minimum of its potential, couples to the W and Z bosons and to all fermions except neutrinos. In either case, the strength of its coupling is proportional to the particle's mass. For this reason, the Higgs boson decays predominantly to the heaviest pair of particles that is kinematically allowed given the Higgs mass. Even though the Higgs mass was not predicted by the theory, it was generally believed that it should be on the "electroweak" scale, that is above 10 and below 1000 GeV. The Feynman diagrams for the dominant Higgs boson decay processes are shown in Fig. 4.1.

4.1.1 *Higgs boson decay to fermions*

The vertex factor in the diagram describing Higgs boson interaction with the fermions is $ig_f = i\frac{m_f\sqrt{2}}{v}$. Let us evaluate the partial width of such a decay. The Higgs boson has zero spin, and couples to right-handed fermions and left-handed antifermions adding up to

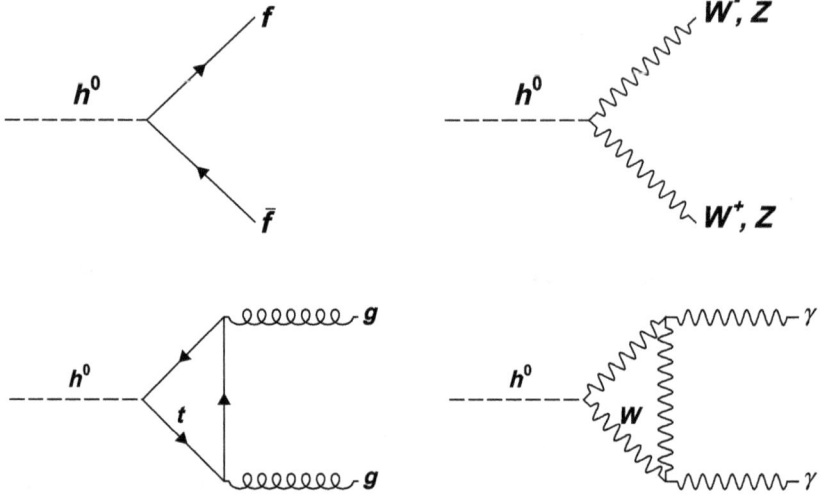

Figure 4.1: The Feynman diagrams for the dominant Higgs boson decay processes.

$s = 1$ state, hence to conserve the angular momentum, the decay must proceed in the p-wave. The matrix element is given by

$$i\mathcal{M} = ig_f \bar{f}_L f_R. \tag{4.1}$$

Summing over the final helicity states similarly to Section 2.8, we get for the matrix element squared

$$\overline{|\mathcal{M}|^2} = \frac{\beta^2 m_f^2 m_h^2}{v^2}, \tag{4.2}$$

where β is the relativistic factor for fermions. Hence, the decay width is given by

$$\Gamma_{h \to f\bar{f}} = \frac{\beta^3 m_f^2 m_h}{8\pi v^2}, \tag{4.3}$$

where we used Eq. (A.17). This formula describes well the Higgs boson decay to leptons.

For the Higgs boson's decays to quarks, QCD corrections are important. First, we need to include a color factor of 3 and corrections due to hard gluon emission, which together yield a factor of

Table 4.1: Quark running masses evaluated at the Higgs boson mass scale.

Quark	Mass (MeV)
Up	1.5
Down	3
Strange	60
Charm	700
Bottom	2,800

1.24. Second, the running quark masses must be evaluated at the Higgs boson mass scale. These masses are given in Table 4.1.

4.1.2 *Higgs boson decay to gauge bosons*

The Feynman diagram for Higgs bosons decays to a pair of gauge bosons $V = W, Z$ is shown in Fig. 4.1 top right. The vertex factor for this diagram is $2i\frac{m_V^2}{v}g_{\mu\nu}$, where m_V is the mass of the gauge boson (consult Table 3.1). The spin-zero Higgs boson can decay to a pair of spin-one bosons in s-wave, given that the gauge bosons have the opposite helicities ($s_1 = -s_2$, described by polarization vectors ϵ_1 and $\epsilon_2 = -\epsilon_1$) and four-momenta k_1 and k_2. The matrix element for this process is given by

$$i\mathcal{M} = 2i\frac{m_V^2}{v}g_{\mu\nu}\epsilon_1^{*\mu}\epsilon_2^{\nu}. \tag{4.4}$$

In the center-of-mass system of the Higgs boson, the energy of each vector boson is equal to half the Higgs boson mass $E_V = m_h/2$, and the momenta are opposite in direction and have the magnitude of $|\vec{k_1}| = |\vec{k_2}| = \sqrt{E_V^2 - m_V^2}$. Using Eq. (A.45), we get for $j = 1, 2$

$$\sum_s \epsilon_j^*\epsilon_j = g_{\mu\nu} - \frac{k_j^\mu k_j^\nu}{m_V^2}. \tag{4.5}$$

Then, using $v = 2m_W/g$, we get

$$\overline{|\mathcal{M}|^2} = \frac{g^2 m_h^4}{4m_W^2} \left(1 - x + \frac{3}{4}x^2\right), \tag{4.6}$$

where $x = \frac{4m_V^2}{m_h^2}$. Using Eq. (A.17), we get the following expression for the partial decay width:

$$\Gamma_{h\to VV} = \frac{\beta g^2 m_h^3 \delta_V}{64\pi m_W^2} \left(1 - x + \frac{3}{4}x^2\right). \tag{4.7}$$

The factor δ_V arises for Z bosons because they are identical. Thus, the integration is performed only over half of the solid angle. As a result $\delta_W = 1$ and $\delta_Z = 1/2$.

It is interesting that in the heavy Higgs boson limit the partial decay width to gauge bosons grows as m_h^3. This behavior can be observed in Fig. 4.2 which shows the total width as predicted in the standard model as a function of the Higgs mass. Should the Higgs boson turn out to be that heavy, it would be a very broad resonance, which is extremely hard to identify experimentally. Luckily, this is not the scenario realized in Nature. It turned out that the Higgs boson has a mass smaller than twice the W boson mass, and for this reason the gauge bosons must be described by off-mass shell propagators.

Another feature that is interesting to note is that in the summation over polarization states, the second term dominates in the heavy Higgs boson case, and thus there is an enhancement of the gauge boson production with longitudinal polarization. For Z boson, it has the following form:

$$\sum_s \epsilon_j^* \epsilon_j \to -\frac{m_h^2}{2m_Z^2} = -\frac{2\lambda}{g^2 + g'^2}. \tag{4.8}$$

Gauge bosons acquire their longitudinal degree of freedom from the Goldstone bosons of the Higgs field. Hence, in the very heavy Higgs boson limit their coupling to the Higgs field behaves like Higgs self-coupling, i.e., is proportional to λ.

4.1.3 *Higgs boson decay branching ratios as a function of its mass*

For low masses ($m_h < 160$ GeV), the Higgs decays predominantly into a pair of b quarks, with subdominant decays into a pair of charm quarks and τ leptons. Long-lived b quarks are identifiable using a secondary decay vertex, which is sufficiently displaced (typically a few millimeters) from the primary collision vertex. A low-mass Higgs has a small enough width to be identified as a narrow resonance (see Fig. 4.2). Yet, the overwhelmingly abundant continuum pair production of b quarks and insufficient energy resolution of hadronic calorimeters make the identification of the leading Higgs boson decay to a pair of b quarks very challenging.

For intermediate range of masses ($160 < m_h < 350$ GeV), the leading decay channels are into a pair of W or Z bosons. Leptons from Z boson decay are readily identifiable and their momenta are well measured, which makes the all-leptonic ZZ mode the golden channel for Higgs boson searches. The W boson decays into a lepton and a neutrino, which is typically not detected, thus the WW decay mode is more challenging experimentally.

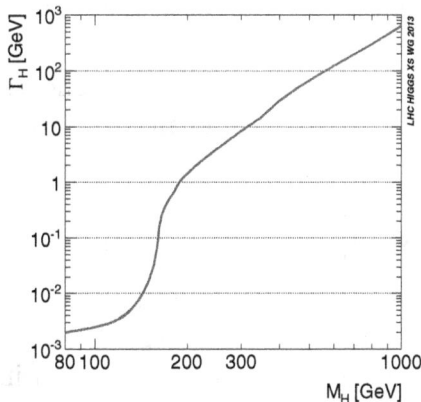

Figure 4.2: The standard model Higgs boson total width, as a function of its mass. For $m_h = 125$ GeV, $\Gamma_h = 4$ MeV.

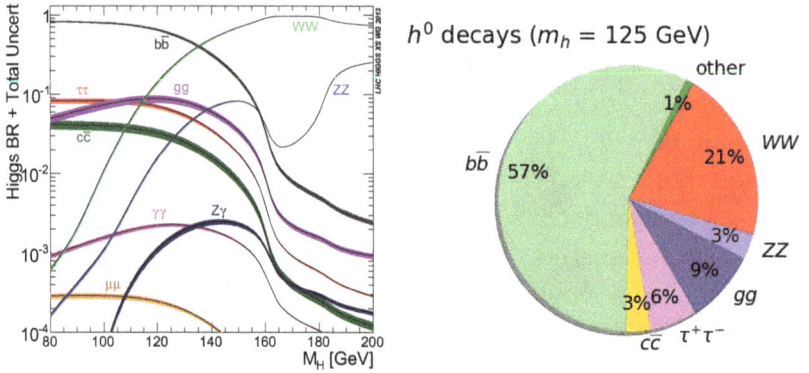

Figure 4.3: Higgs boson branching decay ratios as a function of its mass (left) and pie-chart with dominant decays for $m_h = 125$ GeV (right).

A heavy Higgs boson ($m_h > 350$ GeV) decays predominantly into a pair of top quarks. Each top quark then decays into a b quark and a W boson, which makes the $t\bar{t}$ signature well-distinguished. Yet, the $t\bar{t}$ invariant mass resolution is not so good, making the identification of the Higgs boson over the continuum $t\bar{t}$ production quite challenging.

A couple of loop diagrams are worth mentioning: $h \to \gamma\gamma$ and $h \to \gamma Z$. These processes are mediated by loops, where their main contributors are top quarks and W bosons. Even though branching ratios for these decays are small, superior mass resolution allows for these channels to be well-identified. The Higgs boson decay branching fraction for the range of masses from 80 to 200 GeV is shown in Fig. 4.3.

4.2 Higgs Production in e^+e^- Collisions

The electron's mass is very small, and so is its coupling to Higgs boson. Thus, Higgs boson production in e^+e^- annihilation is incredibly rare. Instead, Higgs boson production in association with a Z boson, mediated by an off-shell Z boson, shown in Fig. 4.4 is the main production mode.

The Large Electron-Positron (LEP) collider operated between 1989 and 2000, in the same tunnel that today hosts the LHC at

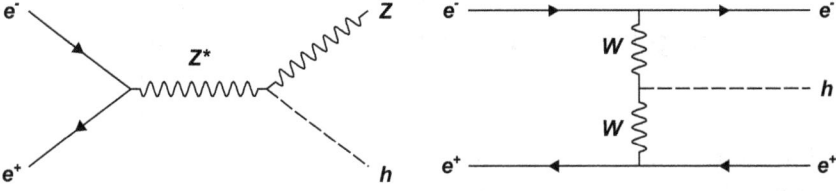

Figure 4.4: Higgs boson production in e^+e^- collisions. Left: "Higgsstrahlung". Right: Vector boson fusion.

CERN. Before LEP, very weak experimental constraints on the Higgs boson mass existed ($m_h > 14$ MeV, coming from nucleon–nucleus scattering). During the LEP I era (up to 1995), the machine operated at $\sqrt{s} \approx m_Z = 91$ GeV, so Higgs production could only be probed by $e^+e^- \to Z \to Z^*H$ with on-shell Z production. Thus, LEP I was only able to exclude m_h up to 64 GeV. After 1995, LEP II operated with $\sqrt{s}$ from 130 to 206 GeV, past the thresholds for WW and ZZ production (161 and 183 GeV, respectively), and this increased the sensitivity to Higgs boson masses up to 115 GeV. The mass reach for the Higgs discovery in e^+e^- machines via Higgsstrahlung is $2E_{\text{beam}} - m_Z$, where E_{beam} is the beam's energy ($\sqrt{s} = 2E_{\text{beam}}$ for head-on collisions).

Due to the low background environment of an e^+e^- collider, the leading Higgs boson decay channel to a b quark pair can be used very effectively by detectors with good vertex resolution and track reconstruction. The searches can be divided based on the Z boson decays (see Fig. 2.12). The dominant final state is the "4 jet" signature: $Zh \to q\bar{q}b\bar{b}$, followed by the "missing energy" channel: $Zh \to \nu\bar{\nu}b\bar{b}$, and the "lepton channel": $Zh \to \ell^+\ell^-b\bar{b}$ (including τ leptons). The difference with the LEP I searches, however, is that now there are two irreducible backgrounds in diboson production, specially ZZ. This spurred the creation of innovative searches fitting the two (non-b) fermions to the Z mass, the development of the first multivariate analyses which achieved signal-to-noise ratios from 1/1 up to 2/1, and statistical techniques still in use today.

In early 2000, the LEP II collider at CERN was leading the search for Higgs boson (see Box 4.1). The name of the game was to get to as high energy as possible. The energy of electron beams is limited by the synchrotron radiation, which is proportional to the fourth power of the relativistic factor γ. LEP II accelerator physicists reached almost impossible beam energies of 104.5 GeV, by adding 288 accelerating superconducting RF cavities to the existing 128 room-temperature copper cavities, which gave an accelerating gradient of 3,630 MV per turn, and fine-tuning every possible step in the machine. Most of the data that year were recorded at $\sqrt{s} = 206$ GeV, but around 7 pb^{-1} at 209 GeV increased the reach to Higgs boson masses slightly above 115 GeV.

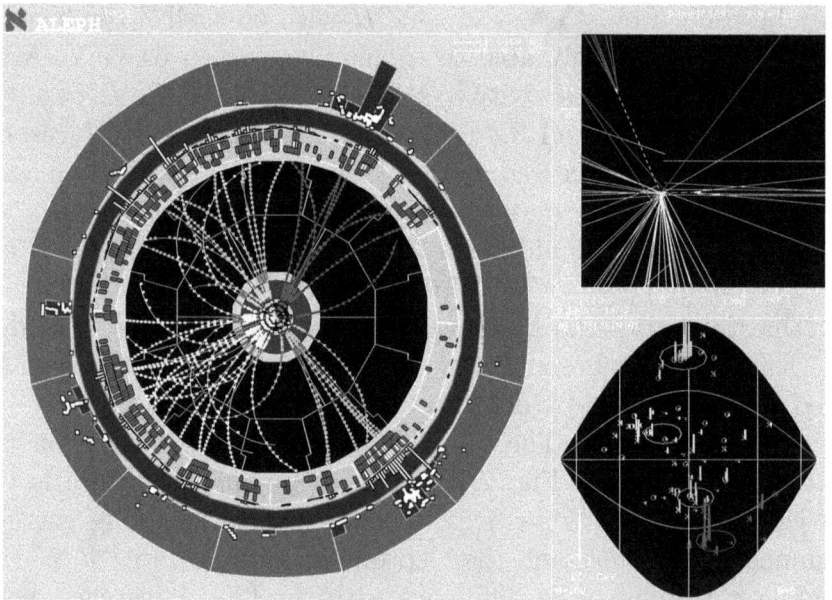

Figure 4.5: Display of a Higgs boson candidate event in the 4-jet category ($Zh \to q\bar{q}b\bar{b}$), recorded by the ALEPH experiment. The displaced secondary vertices of the two b-tagged jets (the red and green showers of tracks) can be seen in the zoomed-in xy view on the upper-right.

Box 4.1 Higgs Searches at LEP II

By the spring of 2000, no Higgs-like events were observed at LEP; and Higgs boson existence in the mass range below 108 GeV was ruled out. By mid-summer, the ALEPH experiment announced a 4-jet signal-like event, with a reconstructed m_h of 114 GeV (Fig. 4.5). Two other 4-jet events were discovered by ALEPH and L3 joined the excitement with a high-significance $\nu\bar{\nu}b\bar{b}$ event and, in early September, CERN delayed the permanent shutdown until November 3rd, giving the experiments a chance to collect more data. One of the authors was a graduate student taking shifts in the ALEPH experiment that summer (and watching the UEFA European championship matches in the control room when there was a chance), and remembers vividly arriving every day wondering if another event had been discovered the previous day: the experiments deployed fast response teams that were processing and scanning the data almost in real time. In the end, as can be seen in Fig. 4.6 (left), the data are compatible with both hypothesis (background-only or Higgs production at $m_h = 115$ GeV). The combined LEP p-value for the background-only hypothesis is 9% (far from 3×10^{-7}, or 5 standard deviations needed to claim discovery). ALEPH had a 3 sigma excess in the 4-jet channel, compatible with $m_h = 114$ GeV. L3 had a small excess, while OPAL and DELPHI had rather background-like distributions. The combined p-value for the signal+background hypothesis for $m_h = 115$ GeV was 15%. The final combined result excluded Higgs boson existence with a mass below 114.4 GeV at 95% CL, as shown in Fig. 4.6 (right). Once again, this experience demonstrated the importance of having several experiments in the same collider complementing each other's strengths and being able to compare and finally combine their results.

(Continued)

(Continued)

Interestingly, if LEP had been equipped with 100 additional superconducting cavities (here again you see the cost of synchrotron radiation in circular e^+e^- colliders), the collision energy would have reached 220 GeV, which could have probed Higgs boson masses up to 129 GeV! If only...

Though at that time tantalizing hints of a signal were presented to the community, CERN management made the decision to stop LEP II operations to begin the construction of the LHC.

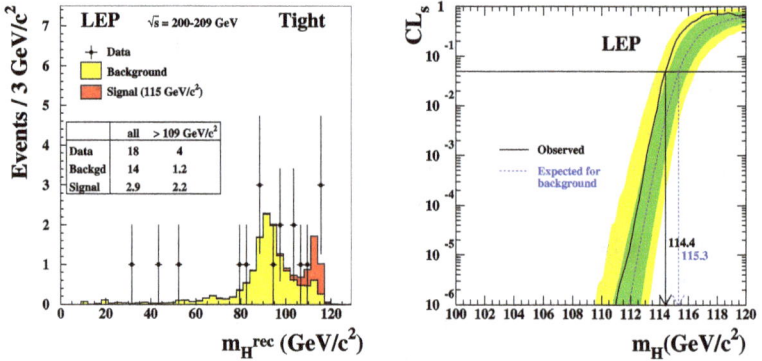

Figure 4.6: (Left) Combination of the reconstructed Higgs mass distributions for all channels and all LEP experiments. (Right) The ratio $CL_s = CL_{s+b}/CL_b$ for the signal plus background hypothesis. Solid line: observation; dashed line: median background expectation. The dark and light shaded bands around the median expected line correspond to the 68% and 95% probability bands. The intersection of the horizontal line for $CL_s = 0.05$ with the observed curve is used to define the 95% confidence level lower bound on the mass of the SM Higgs boson.

4.3 Higgs Production in Proton–Antiproton Collisions

Meanwhile, across the Atlantic Ocean, the Tevatron $p\bar{p}$ collider (see Fig. 4.7) joined the hunt for Higgs boson.

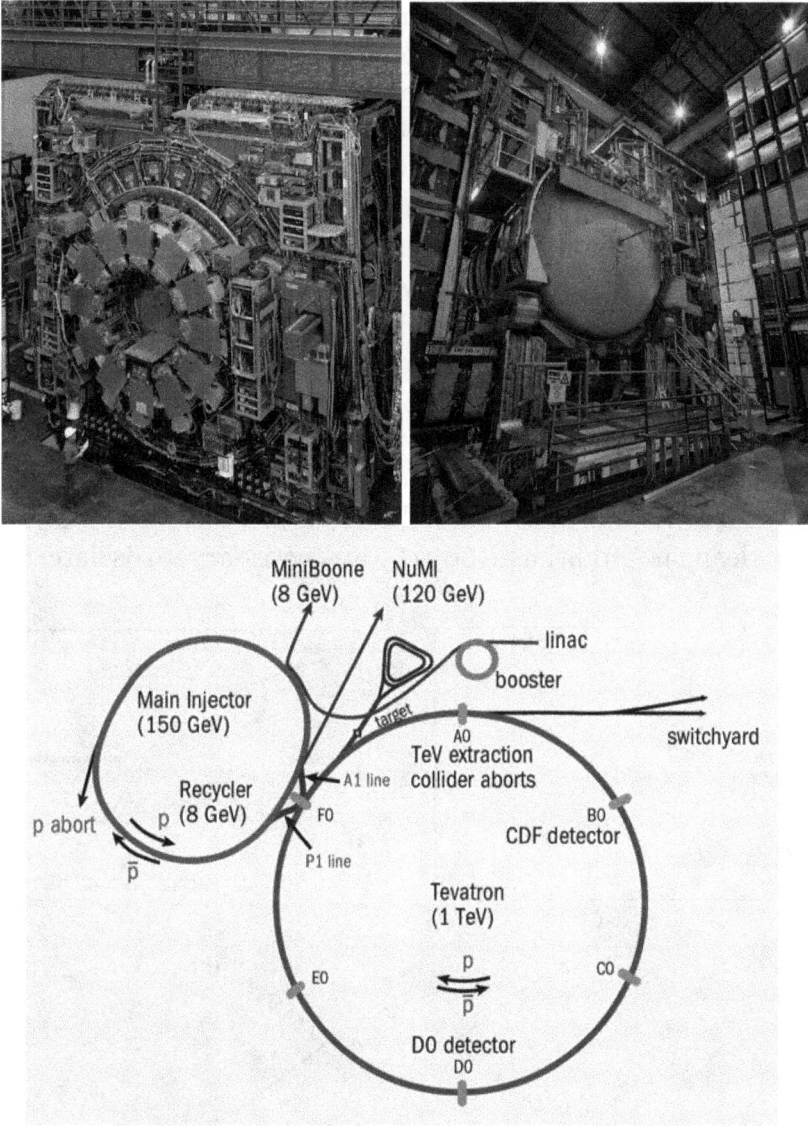

Figure 4.7: The CDF and D0 detectors at the Fermilab Tevatron $p\bar{p}$ accelerator complex, in Batavia, IL.

Cross sections of Higgs boson production in $p\bar{p}$ collisions are shown in Fig. 4.8 for different production channels. The mechanisms of low-mass Higgs production in $p\bar{p}$ collisions is similar to that in e^+e^- machines. Valence quarks from a proton react with valence antiquarks from an antiproton producing virtual W or Z bosons, which decay into Higgs and a real W or Z boson.

Such production mechanisms are collectively referred to as "associate" production. W bosons were identified using their decays to a lepton and a neutrino, while Z bosons were identified in their dilepton decays, where leptons could be either charged, or neutrinos, resulting in an apparent momentum imbalance. Higgs boson was identified using its decays to $b\bar{b}$ pairs.

Intermediate-mass Higgs bosons were also searched for using decays to WW and ZZ. Since the signatures of these decays are fairly clean, no additional objects are necessary to isolate these

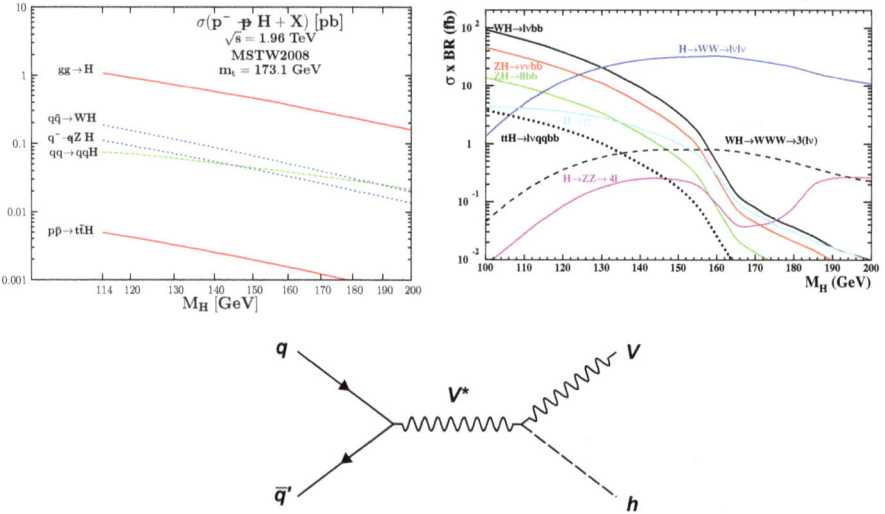

Figure 4.8: (Top left) The Higgs boson production cross section (in pb) at the Tevatron as a function of its mass. (Top right) The production cross section times branching ratio in the most sensitive channels at the Tevatron as a function of mass. The most relevant mechanism, due to the high backgrounds in $gg \rightarrow h$, is the associated production with a V = W, or Z boson, as shown on the Feynman diagram at the bottom.

events from background processes. The Higgs boson by itself is produced via gluon–gluon fusion, mediated by a loop diagram with the top quark being the main contributor to the loop. See Box 4.2 for details of Higgs boson searches at the Tevatron.

Box 4.2 Higgs Searches at Tevatron

In hadron collisions, the constituents of protons and antiprotons are the initial particles of the reaction. The fraction of (anti)proton energy carried by the constituents is not known a priori. Thus, a whole range of center-of-mass energies was probed in the same run of the Tevatron. Yet, the fraction of constituents with a particular value of minimal energy depends on the energy of the beam. After the top center-of-mass energy of 1.96 TeV was reached by 2002, the key parameter was the instantaneous luminosity, which determined the probability for reactions to happen. Increasing the statistics of the sample was crucial in search for rare Higgs boson production processes. A limiting factor in pushing for higher luminosities was the number of antiprotons produced. The Tevatron was taking advantage of the unique antiproton source at Fermilab.

The Tevatron Run II experiments CDF and D0 collected around 10 fb^{-1} of data between 2002 and 2011 at $\sqrt{s} = 1.96$ TeV. The main difference with LEP searches, where both signal and background processes are produced by electroweak interactions, is the overwhelming total inelastic cross section mediated by the strong force, which is more than ten orders of magnitude larger than the Higgs boson production cross section. This means that the Tevatron searches cannot exploit the channels with largest cross section times branching ratio such as gluon fusion with $h \to b\bar{b}$ or $h \to WW(ZZ) \to q\bar{q}q\bar{q}$ decays, and must rely on final states with electrons, muons or missing energy from

(Continued)

(*Continued*)

neutrinos. In addition, due to the huge QCD multijet cross section, high instantaneous luminosity of 4×10^{32} cm^{-2} s^{-1}, and the short time interval between bunch crossings of 396 ns (which implies that reading out an event makes it impossible for the detector to be ready for the next collision), a stringent selection must be implemented in real time to ensure as many as possible signal events are recorded. The irreducible backgrounds are also more numerous and harder to model than at LEP: specially Z/W+jets, which involves the radiation of several jets beyond those at leading order. The excellent performance of the silicon detectors was crucial in achieving highly advanced b-tagging algorithms and good resolution in the invariant $b\bar{b}$ mass to reconstruct the Higgs boson. The use of neural networks and boosted decision trees in most analyses was generalized to increase the sensitivity.

As shown in Fig. 4.8, the channels with highest sensitivity for m_h <130 GeV are $Wh \to \ell\bar{\nu}b\bar{b}$, $Zh \to \nu\bar{\nu}b\bar{b}$ and $Zh \to \ell\ell b\bar{b}$. For $m_h > 130$ GeV, $h \to WW \to \ell^-\bar{\nu}\ell^+\nu$ is the most sensitive channel. These typically have a sensitivity (expressed in terms of expected limits) of around 3–5 times the standard model cross section at 125 GeV for each individual channel, and a combined sensitivity of around $2\sigma_{\mathrm{SM}}$. CDF and D0 launched a huge program to cover many other final states, which contribute between 10% and 20% to the total sensitivity, but which individually may only reach a limit of 7 times σ_{SM}. These include $h \to \gamma\gamma$, $h \to ZZ \to 4\ell$, $h \to \tau\tau$, $h \to WW \to \ell\nu\, q\bar{q}$, $Vh \to q\bar{q}\,b\bar{b}$ and vector boson fusion with $h \to b\bar{b}$, and $t\bar{t}h \to t\bar{t}b\bar{b}$.

Toward the end of its run, the Tevatron was able to exclude a range of available intermediate Higgs boson masses, using WW and ZZ decay modes, and got very close to

(*Continued*)

observing the signal of the low mass Higgs boson (Fig. 4.9). By 2012, Tevatron and LEP searches excluded the majority of allowed Higgs boson masses as demonstrated in Fig. 4.10. The LHC adopted many of the successful strategies developed in CDF and D0: exploiting many different channels with different sensitivities, algorithms to identify b quarks, and the experience of deploying multivariate techniques and combining the final results.

4.4 Higgs Production in Proton–Proton Collisions

Yet, it is the LHC that is rightfully credited with the Higgs boson discovery. Unlike the Tevatron, the LHC is a proton–proton collider. This means that antiquarks come only from the "sea" of virtual quarks and gluons, making them far less energetic. As a result, the Higgs associated production is suppressed compared to the Tevatron. The leading production channels for Higgs boson are via gluon fusion and vector boson fusion (VBF), shown in Fig. 4.11. In the VBF process, two gauge bosons are "shaken off" valence

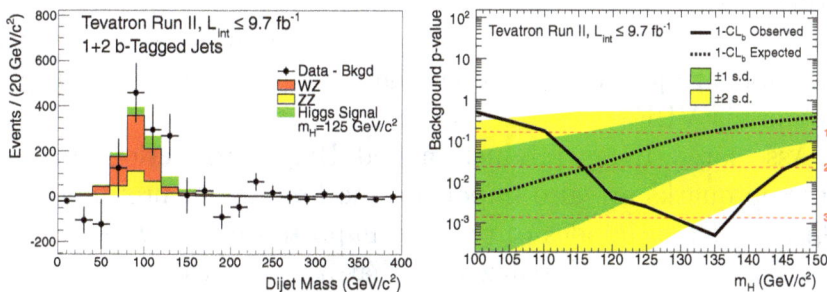

Figure 4.9: (Left) Background-subtracted distribution of the reconstructed dijet mass, summed over all $h \to b\bar{b}$ channels. (Right) Local p-value for the background-only hypothesis as function of Higgs mass for the $h \to b\bar{b}$ search.

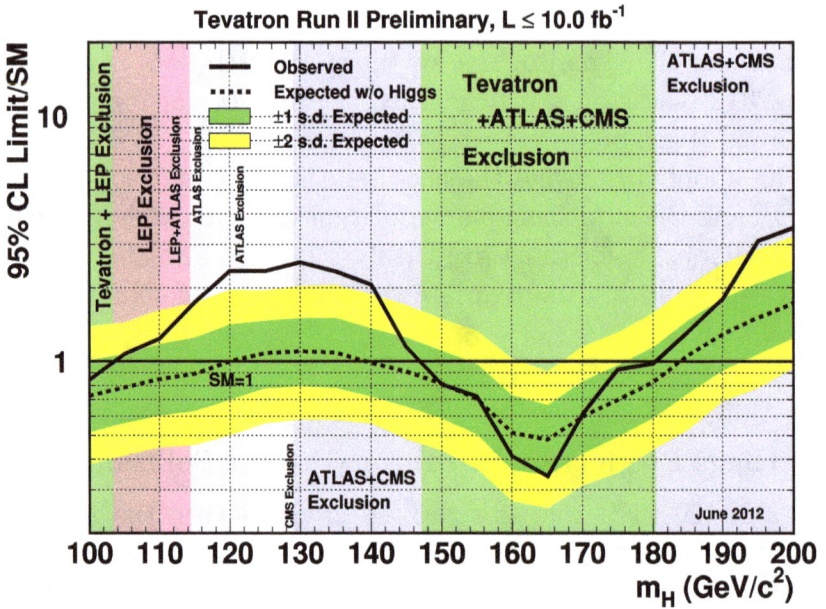

Figure 4.10: The 95% CL exclusion Higgs boson as a function of its mass as of July 2012, just prior to the discovery announcement, including full LEP and Tevatron dataset, as well as CMS and ATLAS results made public at that time.

quarks and fuse to produce a Higgs boson. The characteristic signature of this process is two energetic forward jets produced in a hadronization process of the remnants of the initial state protons. Thus, forward jet tagging was crucial to identification of Higgs production via VBF.

Associated Higgs production and Higgs production together with top quarks are also important mechanisms in proton–proton collisions. Decay modes of primary importance were $h \to \gamma\gamma$ and $h \to Z^{(*)}Z \to 4\ell$. The realization that $h \to \gamma\gamma$ decay mode could be the discovery channel was not trivial at all. The branching ratio of this decay is at the order of 10^{-3}. Yet, with excellent resolution of electromagnetic calorimeters of LHC experiments, a very

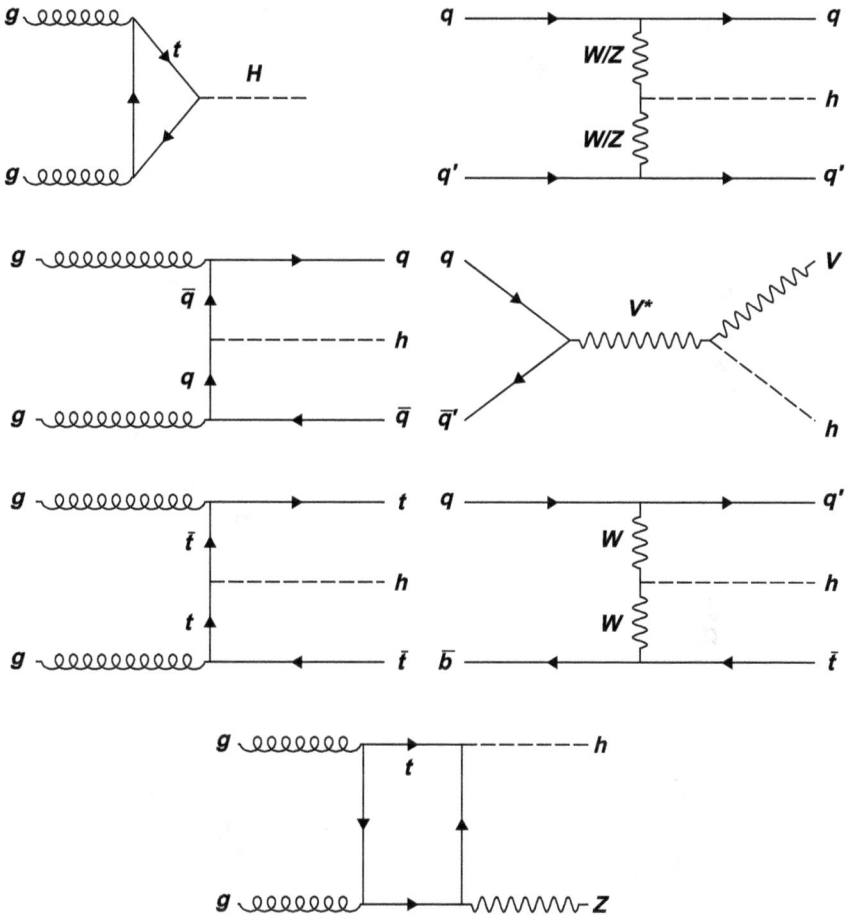

Figure 4.11: Main leading order Feynman diagrams contributing to the Higgs boson production in proton–proton colliders.

narrow resonance can be observed over a large continuum background of diphoton production. The total cross sections and cross sections times branching ratios as predicted by the standard model are shown in Fig. 4.12. However, a possibility that the Higgs boson does not couple to the fermions was also considered, as discussed in Box 4.3.

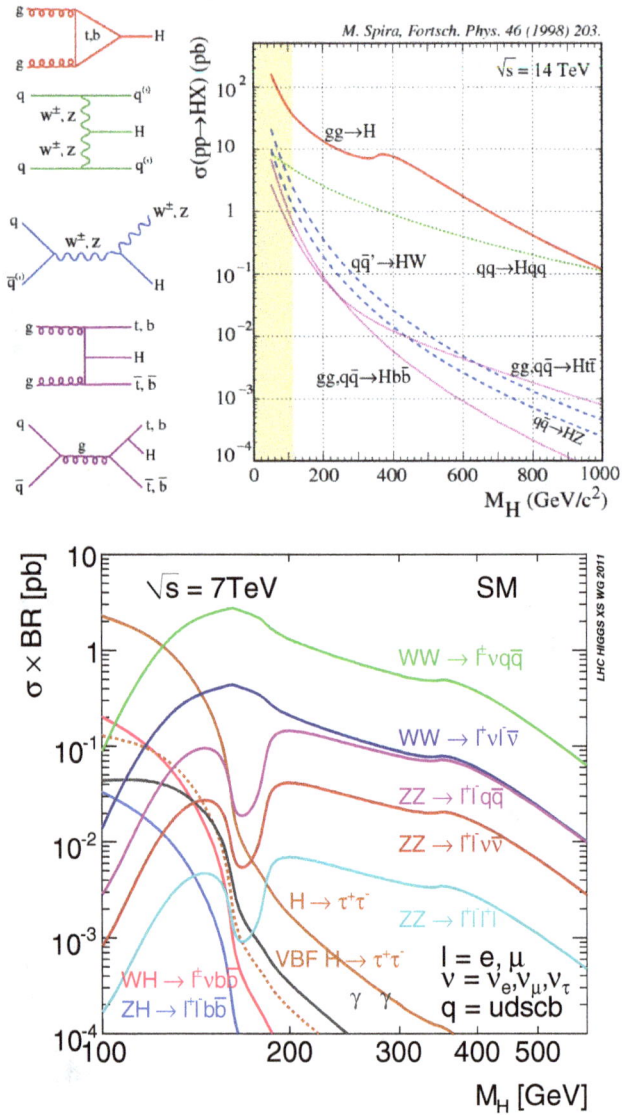

Figure 4.12: Higgs boson production cross section as a function of its mass at the LHC. The diagrams are for Gluon fusion (ggF), vector boson fusion (VBF), qqh, Higgsstrahlung or associated Higgs production (with a gauge boson V), $t\bar{t}h$, tqh and box Zh production. The bottom plot shows some of the main decay channels available at the 7 TeV LHC.

Box 4.3 Fermiophobic Higgs

While generating gauge boson masses is the Higgs-field day job, giving fermions their masses is its extracurricular activity. These Lagrangian terms (as shown in Eq. (3.26)) are introduced ad hoc, there is no explanation for the origin of the Yukawa couplings g_f, whose values vary by six orders of magnitude. The number of free parameters is not reduced by this construction and the unexplained pattern of fermionic masses is replaced by an unexplained pattern of fermion Yukawa couplings. Due to the fact that the Higgs coupling to fermions was added ad hoc, i.e., by hand, there are models that contain no such contributions to the Lagrangian. In these models, the scalar field couples to the gauge bosons, just as described above, but there is no interaction with the fermionic fields, hence this scalar field is referred to as the *fermiophobic* Higgs. Such possibility was an important consideration during the Higgs boson search, since not all channels for production and decay are present in this scenario. Moreover, the absence of the fermionic decay modes alters the branching ratio for the other modes.

In the fermiophobic Higgs scenario, the decays to fermions are not possible, hence for low mass Higgs the leading decay channel is $\gamma\gamma$ mediated solely by the W boson "running" in the loop, since there is no coupling to top quarks. Intermediate and high mass scenarios become indistinguishable, since there is no top-antitop decay option. The associated Higgs production, which is the most significant channel for Higgs searches at LEP II and Tevatron colliders, as well as VBF production at LHC, are available in a fermiophobic Higgs scenario. Gluon fusion and *tth* production mechanisms, on the other hand, are closed to fermiophobic Higgs. The possibility of the fermiophobic scenario was a consideration in the design of the LHC experiments.

Suggested Reading for Chapter 4

[1] LHC Higgs Cross Section Working Group Collaboration. "Handbook of LHC Higgs Cross Sections: 3. Higgs Properties". *CERN Yellow Report* (July 2013), arXiv: 1307.1347 [hep-ph].

[2] LEP Working Group for Higgs boson searches, ALEPH, DELPHI, L3, OPAL Collaboration. "Search for the standard model Higgs boson at LEP". *Phys. Lett. B* 565 (2003), pp. 61–75. arXiv: hep-ex/0306033.

[3] Julien Baglio and Abdelhak Djouadi. "Predictions for Higgs production at the Tevatron and the associated uncertainties". *JHEP* 10 (2010), p. 064. arXiv: 1003.4266 [hep-ph].

[4] S. Dittmaier and M. Schumacher. "The Higgs Boson in the Standard Model — From LEP to LHC: Expectations, Searches, and Discovery of a Candidate". *Prog. Part. Nucl. Phys.* 70 (2013), pp. 1–54. arXiv: 1211.4828 [hep-ph].

[5] CDF, D0 Collaboration. "Evidence for a particle produced in association with weak bosons and decaying to a bottom-antibottom quark pair in Higgs boson searches at the Tevatron". *Phys. Rev. Lett.* 109 (2012), p. 071804. arXiv: 1207.6436 [hep-ex].

[6] CDF, D0 Collaboration. "Updated Combination of CDF and D0 Searches for Standard Model Higgs Boson Production with up to $10.0\,\text{fb}^{-1}$ of Data". July 2012. arXiv: 1207.0449 [hep-ex].

[7] Michael Spira. "QCD effects in Higgs physics". *Fortsch. Phys.* 46 (1998), pp. 203–284. arXiv: hep-ph/9705337.

[8] LHC Higgs Cross Section Working Group Collaboration. "Handbook of LHC Higgs Cross Sections: 1. Inclusive Observables" (Jan. 2011). arXiv: 1101.0593 [hep-ph].

Chapter 5

Higgs Boson Discovery

5.1 Indirect Constraints and Limits from Direct Searches

Even though theory did not predict the value of the Higgs boson mass, other than indicating that it should be at the electroweak scale, experimental constraints on it existed based on indirect observations. The radiative corrections to W boson mass are sensitive to both Higgs boson and top quark masses, as well as the value of the strong coupling, α_S. This sensitivity comes from the loop diagrams shown in Fig. 5.1. Namely, W boson mass depends quadratically on top quark mass and logarithmically on the Higgs boson mass. Because of the very strong dependence of the W boson mass on the top quark mass, precise measurements of W boson mass from LEP II experiments narrowed down the value of top quark mass to 175 ± 15 GeV before the top quark discovery by the Tevatron in 1995. Once top quark was discovered with a mass of 175 ± 5 GeV, some LEP physicists claimed that it was merely a confirmation of their measurement. By the end of the Tevatron run in 2011, the uncertainty on the top quark mass was reduced to 1 GeV. The combination of the precise measurement of the top quark and W boson masses, together with α_S, put a rather stringent constraint on the Higgs boson mass of less than 160 GeV (see Fig. 5.2 left). LEP II excluded Higgs boson masses below 114 GeV. Moreover, direct searches at the Tevatron, using WW and ZZ channels, excluded a mass

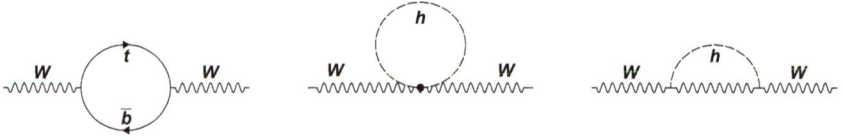

Figure 5.1: The bare mass of the W boson receives radiative corrections from loops involving the top quark and Higgs boson, such as those shown here. Precise measurements of m_W and m_t were able to constrain the mass of the Higgs boson before its discovery because of this sensitivity.

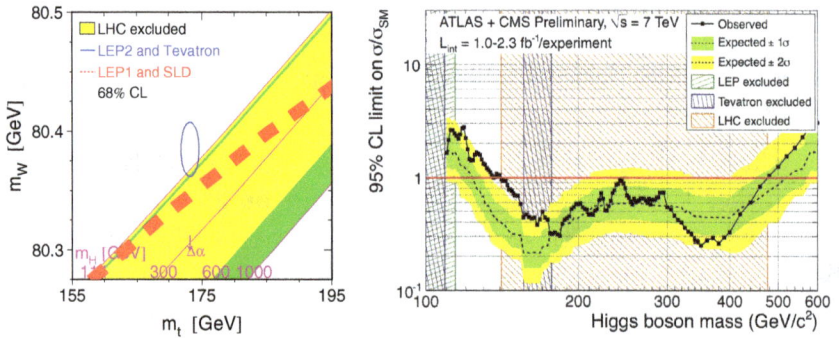

Figure 5.2: (Left) Constraints on the Higgs boson mass from precise measurements of the top quark and W boson masses. (Right) Regions of Higgs boson mass excluded from direct searches by different experiments prior to the discovery.

region from 155 to 185 GeV. Thus, only a rather narrow mass region from 114 to 155 GeV remained (see Fig. 5.2 right). It truly was like finding your car keys in the last pocket you check.

The LEP experiments were optimized for physics at the Z resonance. The Tevatron experiments were largely constructed with top quark search and later top properties' measurements in mind. In both cases, Higgs boson searches were largely opportunistic. For the LHC experiments, on the other hand, Higgs boson searches were their raison d'être. The detectors were designed and constructed with the Higgs boson discovery as the main goal, but the uncertainty in the Higgs boson mass, especially at the design stage of the experiments, demanded that the LHC detectors had to be versatile.

5.2 Higgs Boson Decay to Two Photons

In the 1990s it was realized that the Higgs decay to two photons, despite its miniscule branching ratio, could be the discovery channel. This realization significantly affected the design choices for the LHC experiments. Given the small decay width of the Higgs boson, good energy resolution in the electromagnetic calorimeter became one of the main design goals. Also important was to have high granularity, compactness, operational stability in a magnetic field, radiation hardness, and fast enough response to contribute to the first level trigger. For example, in order to achieve the best possible energy resolution, CMS decided to have a lead-tungstate ($PbWO_4$) scintillating-crystal calorimeter (see Fig. 5.3). Only two factories were producing these crystals — one in Borogoditsk, Russia, and other, the Shanghai Institute of Ceramics (SIC) in China. Both countries were transitioning from totalitarian to market economies

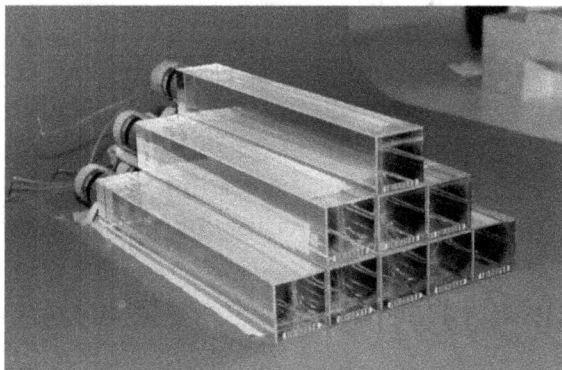

Figure 5.3: Lead-tungstate ($PbWO_4$) crystals for the electromagnetic calorimeter of the CMS detector. In total, CMS has 76,000 lead-tungstate crystals, covering almost 4π steradian, with an energy resolution of 0.5% for high-energy photons.

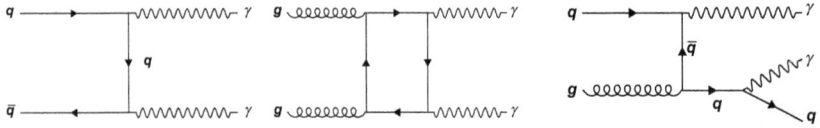

Figure 5.4: Feynman diagrams of the standard model backgrounds in the $h \to \gamma\gamma$ search at the LHC.

at the time. It led to some strong evolution of prices. At the end of the day, this major challenge was overcome using some wise management of the budgeted contingency, but for a while it looked like only the central detector region could be properly instrumented, leaving the endcap region for future upgrades.

Both photons and electrons leave localized energy deposits (clusters) in the electromagnetic calorimeter. To distinguish between the two objects a tracking system is essential. While charged electrons or positrons have a track pointing to the electromagnetic cluster, photons do not. Supreme pointing resolution of the tracker is necessary to make this distinction in the very busy environment of hadron colliders. Still, the standard model continuous production of two photons (some example diagrams responsible for this process are shown in Fig. 5.4) overwhelms diphoton production from Higgs boson decay by orders of magnitude. Only with the employment of advanced analysis methods, experimentalists were able to establish a reliable signal from the Higgs boson. The final diphoton spectrum at the time of discovery is shown in Fig. 5.5. There is a visible bump at around 125 GeV.

5.3 Higgs Boson Decay to a Pair of Z Bosons

Even though a "bump" in the diphoton invariant mass spectrum was observed by both ATLAS and CMS, it would not have been sufficient to convince the physics community that the long sought-after Higgs boson had been discovered. What sealed the deal was the observation in two independent decay channels at the same mass. The second discovery channel was Higgs boson decay to two Z bosons,

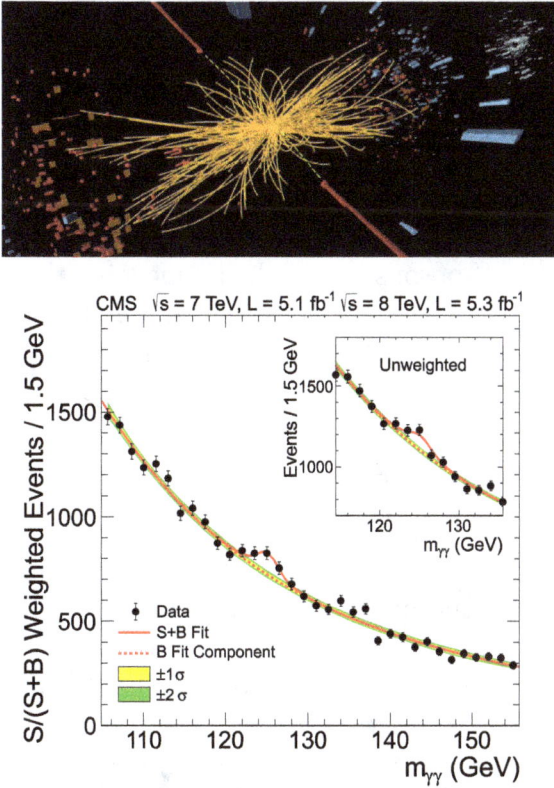

Figure 5.5: $h \to \gamma\gamma$ at the LHC. (Top) Candidate event display in CMS, tracks from charged particles are shown in yellow, and red bars show photon's energy deposition in the electromagnetic calorimeter. (Bottom) Diphoton invariant mass spectrum with each event weighted by the S/(S+B) value of its analysis category. Data points are in black. The lines represent the fitted background and signal, and the coloured bands represent the ± 1 and ± 2 standard deviation uncertainties in the background estimate. The inset shows the central part of the unweighted invariant mass distribution.

with subsequent decay of each Z boson to a pair of leptons, thus resulting in a characteristic signature of 4 leptons. Figure 5.6 shows the diagrams for the background processes in this channel. The probability for such events is very small, resulting in a very clean signal signature. A Higgs boson mass of 125 GeV is not sufficient for it to decay to two Z bosons with a pole mass of $m_Z = 91$ GeV each.

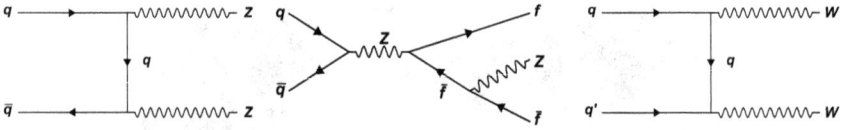

Figure 5.6: Feynman diagrams of the standard model backgrounds in the $h \to ZZ$ search at the LHC.

Thus, one of the Z bosons is "off-shell", meaning that its invariant mass is not equal to the pole mass. Being off-shell suppresses the decay probability since the propagator of the Z boson is given by

$$\frac{1}{p^2 - m_Z^2 + i\Gamma_Z^2}, \tag{5.1}$$

where $p^2 = E^2 - \vec{p}^2$ is the Lorentz four-vector of the Z boson energy-momentum, squared. The further p^2 is from m_Z^2, the smaller is the propagator, and hence the process in question is suppressed. Still, even with this suppression, the cleanliness of the four-lepton signature made this decay channel one of the discovery processes. Figure 5.7 shows the distribution over the invariant mass of four leptons with a clean peak at 125 GeV — same value as the peak in the diphoton invariant mass, as shown in Fig. 5.5.

The discovery of the Higgs boson was announced on July 4, 2012 by ATLAS and CMS experiments simultaneously (see Figs. 5.8 and 5.9).

5.4 Après Discovery

The observation by two independent experiments, in two independent channels, made a very strong case that the discovered particle was indeed the long sought-after Higgs boson. Still the discovery papers stopped short of making such a definitive statement, and claimed the observation of a particle with properties "consistent with that of the scalar boson of the electroweak symmetry breaking". It was necessary to perform a number of properties' measurements before the final statement could be made.

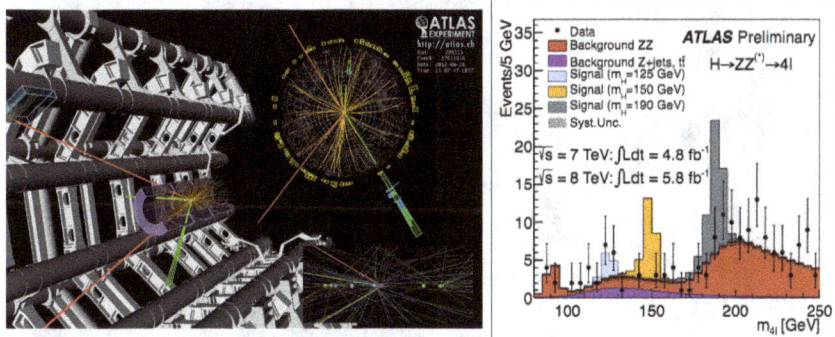

Figure 5.7: $h \to ZZ \to 4\ell$ at the LHC. (Left) Candidate event display for $h \to ZZ \to ee\mu\mu$ in ATLAS. Electron and positron electromagnetic clusters are shown in green. Muon tracks are shown in red. (Right) Invariant mass spectrum of four leptons. Data points are shown in black, the red and purple shaded histograms show the ZZ and Z+jets/$t\bar{t}$ background contributions, and three different expected Higgs signals are plotted for $m_h = 125, 150, 190$ GeV.

Figure 5.8: Higgs discovery announcement at CERN on July 4, 2012: ATLAS spokesperson Fabiola Gianotti.

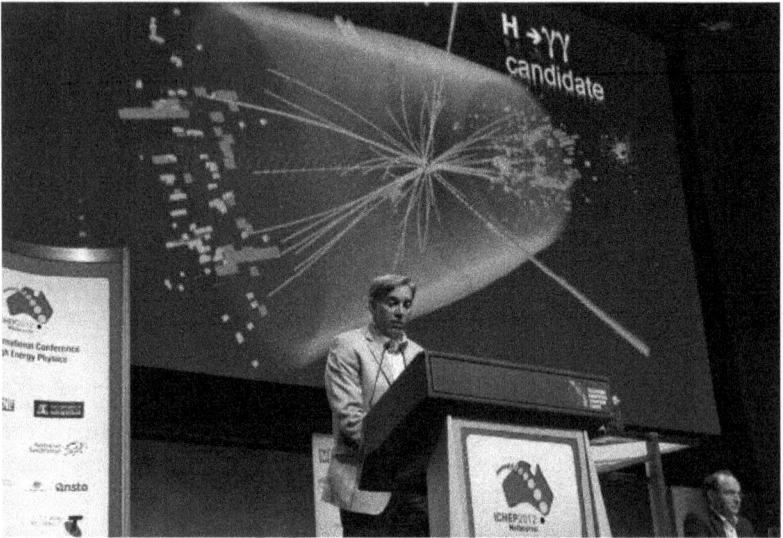

Figure 5.9: CMS spokesperson Joe Incandela presents the Higgs discovery at the ICHEP conference on July 9, 2012 in Melbourne, Australia.

The same observational data allowed to go beyond the claim of the discovery of a new particle. Most importantly, it provided the value of the particle's mass, 125 GeV, not constrained by the theory prior to discovery. This was huge, because now we knew both parameters that described the Higgs potential (e.g., v and μ, or v and λ). Based on these parameters, the strength of the interaction of the Higgs field with any known particle (coupling) and with itself can be predicted. At this point, any additional measurement of these couplings over-constrains the model, and thus becomes a test of its validity, or a search for potential contributions from new physics.

In the following, we discuss the suite of measurements made after the discovery to verify the values of coupling of the Higgs boson to vector bosons, to fermions and hopefully to itself, as well as the Higgs width, spin and parity measurements. All of these are needed to verify that we indeed are dealing with the standard model

Higgs boson and probe for potential deviations from predictions that could open the door to new physics.

Suggested Reading for Chapter 5

[1] ATLAS Collaboration. "Observation of a new particle in the search for the Standard Model Higgs boson with the ATLAS detector at the LHC". *Phys. Lett. B* 716 (2012), pp. 1–29. arXiv: 1207.7214 [hep-ex].

[2] CMS Collaboration. "Observation of a New Boson at a Mass of 125 GeV with the CMS Experiment at the LHC". *Phys. Lett. B* 716 (2012), pp. 30–61. arXiv: 1207.7235 [hep-ex].

[3] CMS Collaboration. "Observation of a New Boson with Mass Near 125 GeV in *pp* Collisions at $\sqrt{s} = 7$ and 8 TeV". *JHEP* 06 (2013), p. 081. arXiv: 1303.4571 [hep-ex].

Chapter 6

Higgs Boson Properties

The standard model predicts the existence of a scalar field, i.e., with spin equal to zero, with positive parity that couples to weak gauge bosons, which acquire mass as a result of this coupling. Couplings to fermions with strengths proportional to the fermion's mass are also postulated. Based on the first experimental observations, it was established that the newly discovered particle can decay to two photons as well as to two Z bosons (Fig. 4.1). Both processes are important in establishing the basic properties of the Higgs boson — its spin, parity, mass and width.

6.1 Spin and Parity: Is it Higgs?

The spin of the newly discovered particle is clearly an integer, since its decay products — photons and Z bosons — have spin 1, which according to quantum mechanical rules cannot add up to a half-integer number.

According to the Landau–Yang theorem, a spin-one particle cannot decay to two photons. Since the newly observed particle did decay into two photons, this rules out the spin-one hypothesis. The remaining options are particles of spin 0 or 2. These two cases result in different angular correlations of the final state particles in ZZ^* decay channel. Another parameter that affects these angular correlations is parity, which can be either positive or negative.

The analysis that sorted out these options utilized the complete information about the angles of the four final state leptons in $h \to ZZ^*$ channel.

The system of four leptons is described by 6 angles. We define the coordinate system such that the z-axis is directed along the beam, which we assume to coincide with the direction of the boost of the system. All decays are symmetric around the beam axis, hence there is no dependence on the azimuthal angle. There are different ways to introduce the other five independent angles. One of them is illustrated in Fig. 6.1. In the rest frame of the Higgs boson the momenta of the two Z bosons add up to zero, so they have to lie on one axis, which forms the polar angle, θ^*, with the z-axis. Together with the beam axis, the Z-boson axis defines the Higgs decay plane. Each Z boson decays to a pair of leptons. Each pair defines another plane, which we will refer to as decay plane of Z_1 and Z_2. The lepton opening angles are called θ_1 and θ_2, where

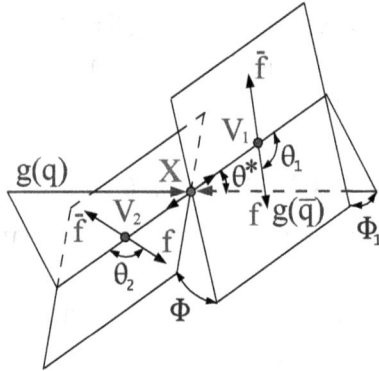

Figure 6.1: Illustration of the production of a particle X in a collision and its decay to two vector bosons: gg or $q\bar{q} \to X \to ZZ$, WW, $Z\gamma$ and $\gamma\gamma$ either with or without sequential decay of each vector boson to a fermion–antifermion pair. The two production angles θ^* and ϕ_1 are shown in the X rest frame and the three decay angles θ_1, θ_2 and ϕ are shown in the V rest frames. Here X stands either for a Higgs boson, an exotic particle, or, in general, the genuine or misidentified VV system, including background.

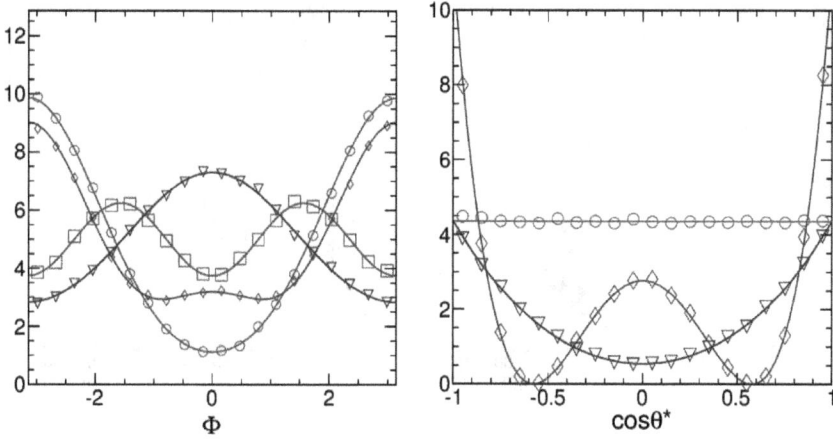

Figure 6.2: Distribution of some of the angular variables from Fig. 6.1. Four signal hypotheses are shown: standard model Higgs boson (red circles), pseudoscalar 0^- (magenta squares), graviton-like tensor with minimal couplings 2^+_m (blue triangles), and tensor with higher-dimension operators 2^+_h (green diamonds). The mass of the resonance is taken to be 125 GeV.

the index refers to the corresponding Z boson. Finally, the angles between the Z_1 and Z_2 decay planes and the Higgs boson decay plane are called ϕ_1 and ϕ_2, respectively, completing the set of five independent angles.

Figure 6.2 shows the distribution over several angles for the tested spin, J and parity, P, hypotheses: $J^P = 0^+$ (standard model), 0^-, 2^+, 2^-. Each of these hypotheses is tested individually against the data. Based on the angular information, a likelihood function L_{JP} is constructed. The test statistic is the ratio of likelihood functions for the hypothesis under test with the one expected from the standard model. A thousand toy experiments are simulated using the angular distributions of a particular hypothesis. As shown in Fig. 6.3, less than 1% of these experiments have higher values of the test statistic observed in data, allowing to exclude the corresponding hypothesis at 99% C.L. By excluding the alternative hypotheses, this important analysis confirmed that the

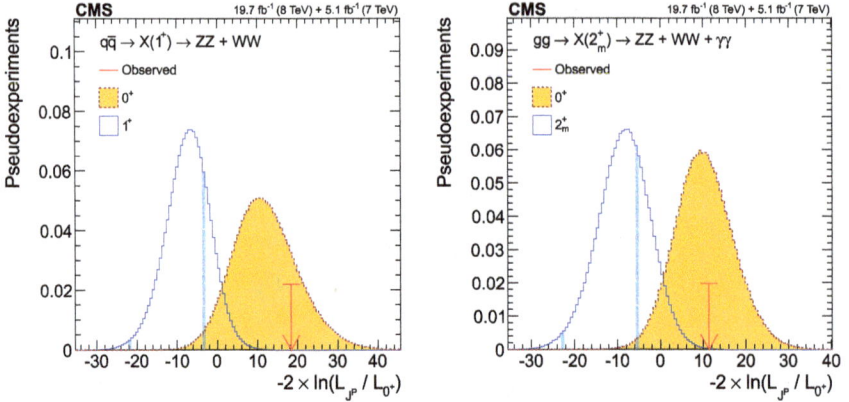

Figure 6.3: Distributions of the test statistic for the hypothesis (left) $q\bar{q} \rightarrow X(1^+)$ and (right) $gg \rightarrow X(2_m^+)$ tested against the standard model Higgs boson hypothesis (0^+). The expectation for the standard model Higgs boson is represented by the yellow histogram and the alternative J^P hypothesis by the blue histogram. The red arrow indicates the observed value.

discovered particle has zero spin and positive parity, as predicted by the theory.

6.2 Mass

The Higgs boson mass is typically evaluated as the peak value in the invariant mass distribution of the two photons or the four leptons. Leptons in this case are electrons and muons; short-lived tau-leptons, which are reconstructed via their decay products, are not used in these analyses. The excellent energy resolution of the electromagnetic calorimeters of both ATLAS and CMS ensures good energy measurements for photons and electrons. Muons are reconstructed as a combination of tracks left in the muon system and the tracks from the central tracking system. Charged particle momenta are inferred from the curvature in magnetic field. Large lever arm ensures good momentum resolutions for muons. Additionally, since one of the Z bosons is on-mass shell, a mass constraint can be imposed on the momenta of a pair of opposite sign leptons that is closer to the Z boson mass, improving the momentum resolution for leptons even further. As a result, even with the very first data it was possible to measure the mass of the newly discovered particle with a very good precision $126.0 \pm 0.4(\text{stat}) \pm 0.4(\text{sys})$ GeV for ATLAS, and $126.2 \pm 0.4(\text{stat}) \pm 0.5(\text{sys})$ GeV for CMS. With more data for the full Run I in 2011 and 2012, and better calibrated energy resolution for the diphoton and four-lepton channels, the ATLAS and CMS combined measurement became $m_h = 125.09 \pm 0.21(\text{stat}) \pm 0.11(\text{sys})$ GeV.

Figure 6.4 shows the invariant mass distribution of the two photons and the four leptons. The $h \rightarrow \gamma\gamma$ analysis performs an unbinned maximum likelihood fit to extract the mass and the signal strength, and the $h \rightarrow ZZ^*$ analysis performs a two-dimensional fit in the four-lepton mass and a Boosted Decision Tree (BDT) discriminant designed to separate Higgs boson events from the ZZ^* background.

These mass measurements are of particular importance since the Higgs boson mass was not predicted by the theory, yet its value affects a number of other theoretical predictions, in particular the Higgs couplings to other particles and as a result Higgs production and decay rates. The Higgs boson together with the top quark contributes to the radiative corrections to the W boson mass. Prior to the Higgs boson discovery, the precise measurements of W boson and top quark masses were used to constrain the expected values of the Higgs boson mass. After the discovery, the model became overconstrained, which allows for the tests of its validity.

6.3 Width

Given the value of the Higgs boson mass and assuming the standard model predicted couplings, it is possible to calculate the decay probabilities of the Higgs boson in all allowed decay channels. Hence, the total width of the Higgs boson decay (Γ_h) is also known within the framework of the standard model: it is expected to be

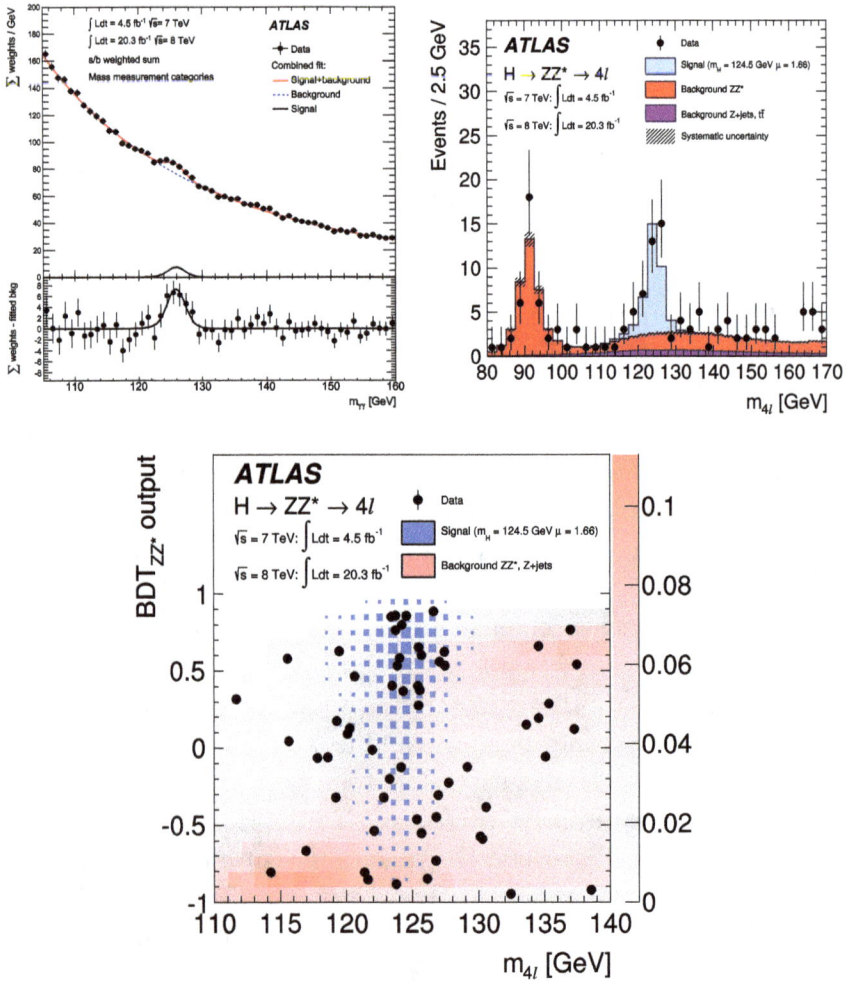

Figure 6.4: Invariant mass distribution in the ATLAS data for the 7 and 8 TeV samples combined, for (left) the $h \to \gamma\gamma$ analysis with weighted data points with errors, and the result of the simultaneous fit to all categories; and (right) for the $h \to ZZ^* \to 4\ell$ analysis in the range of 80–170 GeV (including the signal for $m_h = 124.5$ GeV normalized to the measured signal strength). The final plot (bottom) shows the distribution of the BDT analysis discriminant as a function of the invariant mass for the selected candidates in the 110–140 GeV range.

around 4 MeV, which is significantly below the experimental resolution on the mass of Higgs boson in both the $\gamma\gamma$ and ZZ channels. Yet, this parameter is of great theoretical importance since any deviations from the standard model predicted value would signal either a contribution of an unknown decay channel, or deviations of couplings from their predicted values. In either case, it is a sign of physics beyond the standard model.

A clever method to experimentally constrain the value of total Higgs boson width was suggested by F. Caola and K. Melnikov using an off-resonance $ZZ \to 4\ell$ production. The dependence of Higgs boson production via gluon fusion mechanism on the invariant mass of the four final state leptons, $m_{4\ell}$, is given by

$$\frac{d\sigma_{gg \to h \to ZZ}}{dm_{4\ell}^2} \sim \frac{g_{ggh}^2 g_{hZZ}^2}{(m_{4\ell}^2 - m_h^2)^2 + m_h^2 \Gamma_h^2}, \tag{6.1}$$

where g_{ggh} and g_{hZZ} are couplings of Higgs boson to gluons (via a loop diagram) and Z bosons, respectively, corresponding to Higgs boson coupling in the initial and final states. In the narrow region of $m_{4\ell}$ around the resonance $|m_{4\ell} - m_h| \sim \Gamma_h$. Assuming that the width is much smaller than Higgs mass $\Gamma_h \ll m_h$, the cross section can be presented as

$$\sigma_{gg \to h \to ZZ^*}^{\text{resonance}} \sim \frac{g_{ggh}^2 g_{hZZ}^2}{m_h \Gamma_h}, \tag{6.2}$$

where we neglect terms of higher order in Γ_h/m_h. At the same time in the production cross section off the resonance, $m_{4\ell} - m_h \gg \Gamma_h$, and near the ZZ production threshold, $m_{ZZ} \gtrsim 2m_Z$, the term dependent on Γ_h can be neglected, yielding

$$\sigma_{gg \to h^* \to ZZ}^{\text{off-resonance}} \sim \frac{g_{ggh}^2 g_{hZZ}^2}{(2m_Z)^2}. \tag{6.3}$$

Thus, the ratio of the rates of four-lepton production on- and off-resonance provides a direct access to the width of the Higgs boson. The caveat here is that we must assume that the couplings g_{ggh}

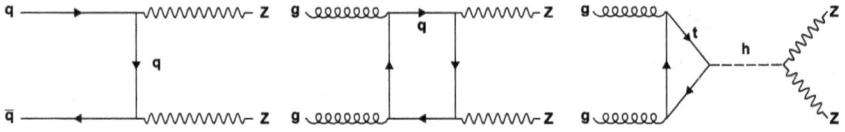

Figure 6.5: Lowest order contributions to the main ZZ production processes: (left) quark-initiated production, $q\bar{q} \to ZZ$, (center) gg background production, $gg \to ZZ$, and (right) Higgs-mediated gg production, $gg \to h \to ZZ$, the signal.

and g_{hZZ} do not change with the invariant mass. While it is not a bad assumption for g_{hZZ}, which is a tree-level interaction in the standard model, it is not a trivial assumption for g_{ggh}, which is an effective coupling, described by loop diagrams dominated by top (and to some extent bottom) quarks. Since the deviation of the Higgs boson width from the standard model prediction implies a contribution of new decay products, if they couple strongly, they might as well contribute to the ggh loop diagram.

Moreover, other mechanisms of the continuum ZZ production must be taken into account. These are $q\bar{q} \to ZZ$, and $gg \to ZZ$, assisted by a quark box diagram (see Fig. 6.5). The latter process interferes with the signal process $gg \to h^* \to ZZ$, since the two processes have the same initial and final state. In the formalism of CMS analysis, the strength of the resonance contribution is constrained to its standard model value, while the rate of the off-resonant production provides the sensitivity to the width of Higgs boson. This analysis led to an upper limit on the Higgs boson width of $\Gamma_h < 22$ MeV at 95% confidence level, which is 5.4 times the value expected in the standard model.

An interesting alternative approach to the evaluation of the width of the Higgs boson utilizes the power of the vertex detectors. The width of a resonance is related to its lifetime, τ_h, through a simple relationship:

$$\Gamma_h = \frac{\hbar}{\tau_h}, \tag{6.4}$$

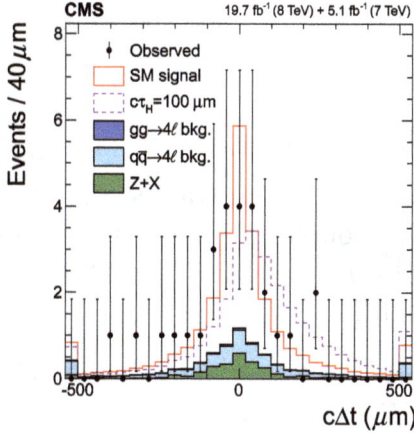

Figure 6.6: The $c\tau$ distribution for signal-like events in the $ZZ \to 4\ell$ channel from CMS. The points with error bars represent the observed data, and the filled histograms stacked on top of each other represent the expected contributions from the standard model backgrounds. Stacked on the total background contribution, the open histograms show the combination of all production mechanisms expected in the standard model for the H boson signal with either the standard model lifetime ($c\tau_h = 48$ fm, beyond the instrumental precision) or $c\tau_h = 100$ μm.

where $\hbar$ is Planck's reduced constant, which is 1 in natural units. Thus, measuring the lifetime, or decay length $c\tau_h$, is equivalent to measuring the width. The distribution of Higgs boson candidates decaying to ZZ and then to four leptons over the decay length is shown in Fig. 6.6. From this, a limit on the lifetime is set at $\tau_h < 0.19$ picoseconds, which corresponds to the limit on the width of $\Gamma_h < 26$ MeV.

The latest most precise constraint on the Higgs boson width is achieved by combining 7, 8 and 13 TeV data in a matrix element analysis based on the kinematic distributions of the four leptons, following the ideas developed to measure the coupling as illustrated in Fig. 6.1, and again including on-shell and off-shell regions. Assuming a coupling structure similar to that in the standard model, the Higgs boson width is constrained to be $3.2^{+2.8}_{-2.2}$ MeV, while the expected constraint based on simulation is $4.1^{+5.0}_{-4.0}$ MeV.

The measured width is in a good agreement with the standard model expectation.

Suggested Reading for Chapter 6

[1] CMS Collaboration. "Constraints on the spin-parity and anomalous HVV couplings of the Higgs boson in proton collisions at 7 and 8 TeV". *Phys. Rev. D* 92.1 (2015), p. 012004. arXiv: 1411.3441 [hep-ex].

[2] Bolognesi, Sara and Gao, Yanyan and Gritsan, Andrei V. and Melnikov, Kirill and Schulze, Markus and Tran, Nhan V. and Whitbeck, Andrew. "On the spin and parity of a single-produced resonance at the LHC". *Phys. Rev. D* 86 (2012), p. 095031

[3] ATLAS Collaboration. "Measurement of the Higgs boson mass from the $H \to \gamma\gamma$ and $H \to ZZ^* \to 4\ell$ channels with the ATLAS detector using $25\,\text{fb}^{-1}$ of pp collision data". *Phys. Rev. D* 90.5 (2014), p. 052004. arXiv: 1406.3827 [hep-ex].

[4] ATLAS, CMS Collaboration. "Combined Measurement of the Higgs Boson Mass in pp Collisions at $\sqrt{s} = 7$ and 8 TeV with the ATLAS and CMS Experiments". *Phys. Rev. Lett.* 114 (2015), p. 191803. arXiv: 1503.07589 [hep-ex].

[5] ATLAS Collaboration. "Measurement of the Higgs boson mass in the $H \to ZZ^* \to 4\ell$ and $H \to \gamma\gamma$ channels with $\sqrt{s} = 13$ TeV pp collisions using the ATLAS detector". *Phys. Lett. B* 784 (2018), pp. 345–366. arXiv: 1806.00242 [hep-ex].

[6] CMS Collaboration. "Constraints on the Higgs boson width from off-shell production and decay to Z-boson pairs". *Phys. Lett. B* 736 (2014), pp. 64–85. arXiv: 1405.3455 [hep-ex].

[7] CMS Collaboration. "Limits on the Higgs boson lifetime and width from its decay to four charged leptons". *Phys. Rev. D* 92.7 (2015), p. 072010. arXiv: 1507.06656 [hep-ex].

[8] CMS Collaboration. "Measurements of the Higgs boson width and anomalous HVV couplings from on-shell and off-shell production in the four-lepton final state". *Phys. Rev. D* 99.11 (2019), p. 112003. arXiv: 1901.00174 [hep-ex].

[9] Caola, Fabrizio and Melnikov, Kirill. "Constraining the Higgs boson width with ZZ production at the LHC". *Phys. Rev. D* 88 (2013), p. 054024.

Chapter 7

Couplings

The Higgs boson discovery was achieved mainly relying on two channels $h \to \gamma\gamma$ and $h \to ZZ^*$. Soon after, several other decay channels were observed. In particular, $h \to WW^*$, $h \to \tau^+\tau^-$ and, somewhat latter, $h \to b\bar{b}$. Different production channels were also isolated, most notably gluon fusion and VBF. The latter process is characterized by two forward energetic jets. Other established processes include associated production, characterized by additional vector boson in the final state, and $t\bar{t}h$ production, where Higgs boson is accompanied by top quark and antiquark. Given the Higgs boson mass, the standard model predicts the rates of each particular process. Thus, each rate measurement is a test of the standard model validity. To quantify the level of agreement, a set of μ parameters, referred to as signal strengths, is introduced. For each process i, the experimentally measured signal strength μ_i is defined as the ratio of the observed cross section $\sigma_i^{\text{observed}}$ to its theoretically predicted value $\sigma_i^{\text{predicted}}$:

$$\mu_i = \frac{\sigma_i^{\text{observed}}}{\sigma_i^{\text{predicted}}} . \tag{7.1}$$

By definition, in the standard model all parameters μ are expected to be 1.

111

7.1 Separating Production Channels

It is clear upon examination of the Higgs boson production diagrams via gluon fusion (ggF) and via VBF that the two provide access to different couplings. The former is sensitive to top and, to some extent, bottom quark couplings via loop contribution, while the later relies on Higgs boson couplings to W and Z bosons. Thus, disentangling the two production mechanisms is highly desirable, yet far from trivial. VBF production results in two additional forward jets, which are typically quite energetic. Not all jets in the forward direction are identified due to detector gaps and inefficiency in the reconstruction (see Fig. 7.1 (left)). At the same time, when the Higgs boson is produced in gluon fusion, there might be additional jets due to initial state radiation, underlying event remnants or multiple interactions. An additional handle to separate the two production mechanisms is to use the transverse momentum of the Higgs boson decay products. Higgs bosons produced via gluon fusion typically have low values for their transverse momentum, while VBF-produced Higgs bosons, recoiling from the two

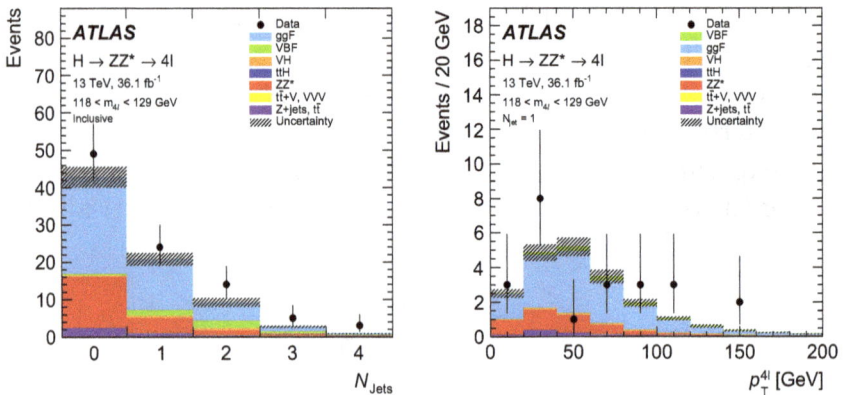

Figure 7.1: Kinematic distributions in $h \to ZZ^* \to 4\ell$ channel for (left) number of additional jets and (right) transverse momentum of the four-lepton system.

energetic jets, tend to have a higher momentum. This is illustrated in Fig. 7.1(right).

Other event characteristics provide some additional discriminating power. To maximally utilize all the available information, events are divided into different categories based on the number and transverse momenta of jets, transverse momentum of the Higgs boson candidate, invariant mass of two jets and the number of leptons. Based on the number of events in each category, the signal strength is measured for gluon fusion, VBF, associated production and $t\bar{t}h$ production as shown in Fig. 7.2.

Using this strategy, the μ parameters were evaluated for several decay and production channels, which as demonstrated by Fig. 7.3 are largely consistent with the prediction of the standard model within their uncertainties.

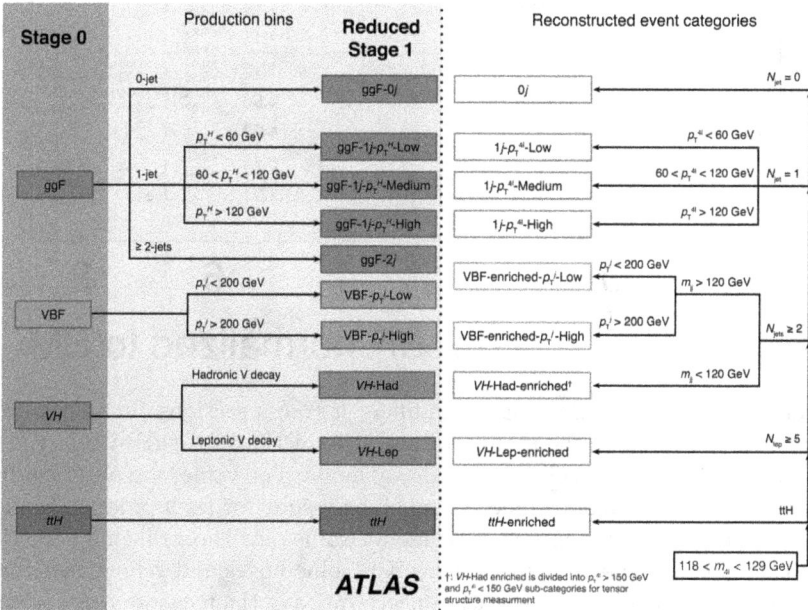

Figure 7.2: Event categorization by production channel for the ATLAS analysis of the $h \to ZZ^* \to 4\ell$ decay.

ATLAS   ⊢━⊣ Total   ☐ Stat.   ▬ Syst.   ▓ SM
$\sqrt{s}$ = 13 TeV, 24.5 - 79.8 fb⁻¹
m_H = 125.09 GeV, $|y_H|$ < 2.5
p_{SM} = 71%

			Total	Stat.	Syst.
ggF	$\gamma\gamma$		0.96	± 0.14	(± 0.11 , $^{+0.09}_{-0.08}$)
	ZZ^*		1.04	$^{+0.16}_{-0.15}$	(± 0.14 , ± 0.06)
	WW^*		1.08	± 0.19	(± 0.11 , ± 0.15)
	$\tau\tau$		0.96	$^{+0.59}_{-0.52}$	($^{+0.37}_{-0.36}$, $^{+0.46}_{-0.38}$)
	comb.		1.04	± 0.09	(± 0.07 , $^{+0.07}_{-0.06}$)
VBF	$\gamma\gamma$		1.39	$^{+0.40}_{-0.35}$	($^{+0.31}_{-0.30}$, $^{+0.26}_{-0.19}$)
	ZZ^*		2.68	$^{+0.98}_{-0.83}$	($^{+0.94}_{-0.81}$, $^{+0.27}_{-0.20}$)
	WW^*		0.59	$^{+0.36}_{-0.35}$	($^{+0.29}_{-0.27}$, ± 0.21)
	$\tau\tau$		1.16	$^{+0.58}_{-0.53}$	($^{+0.42}_{-0.40}$, $^{+0.40}_{-0.35}$)
	$b\bar{b}$		3.01	$^{+1.67}_{-1.61}$	($^{+1.63}_{-1.57}$, $^{+0.39}_{-0.36}$)
	comb.		1.21	$^{+0.24}_{-0.22}$	($^{+0.18}_{-0.17}$, $^{+0.16}_{-0.13}$)
VH	$\gamma\gamma$		1.09	$^{+0.58}_{-0.54}$	($^{+0.53}_{-0.49}$, $^{+0.25}_{-0.22}$)
	ZZ^*		0.68	$^{+1.20}_{-0.78}$	($^{+1.18}_{-0.77}$, $^{+0.18}_{-0.11}$)
	$b\bar{b}$		1.19	$^{+0.27}_{-0.25}$	($^{+0.18}_{-0.17}$, $^{+0.20}_{-0.18}$)
	comb.		1.15	$^{+0.24}_{-0.22}$	(± 0.16 , $^{+0.17}_{-0.16}$)
$t\bar{t}H+tH$	$\gamma\gamma$		1.10	$^{+0.41}_{-0.35}$	($^{+0.36}_{-0.33}$, $^{+0.19}_{-0.14}$)
	VV^*		1.50	$^{+0.59}_{-0.57}$	($^{+0.43}_{-0.42}$, $^{+0.41}_{-0.38}$)
	$\tau\tau$		1.38	$^{+1.13}_{-0.96}$	($^{+0.84}_{-0.76}$, $^{+0.75}_{-0.59}$)
	$b\bar{b}$		0.79	$^{+0.60}_{-0.59}$	(± 0.29 , ± 0.52)
	comb.		1.21	$^{+0.26}_{-0.24}$	(± 0.17 , $^{+0.20}_{-0.18}$)

−2   0   2   4   6   8

$\sigma \times$ BR normalized to SM

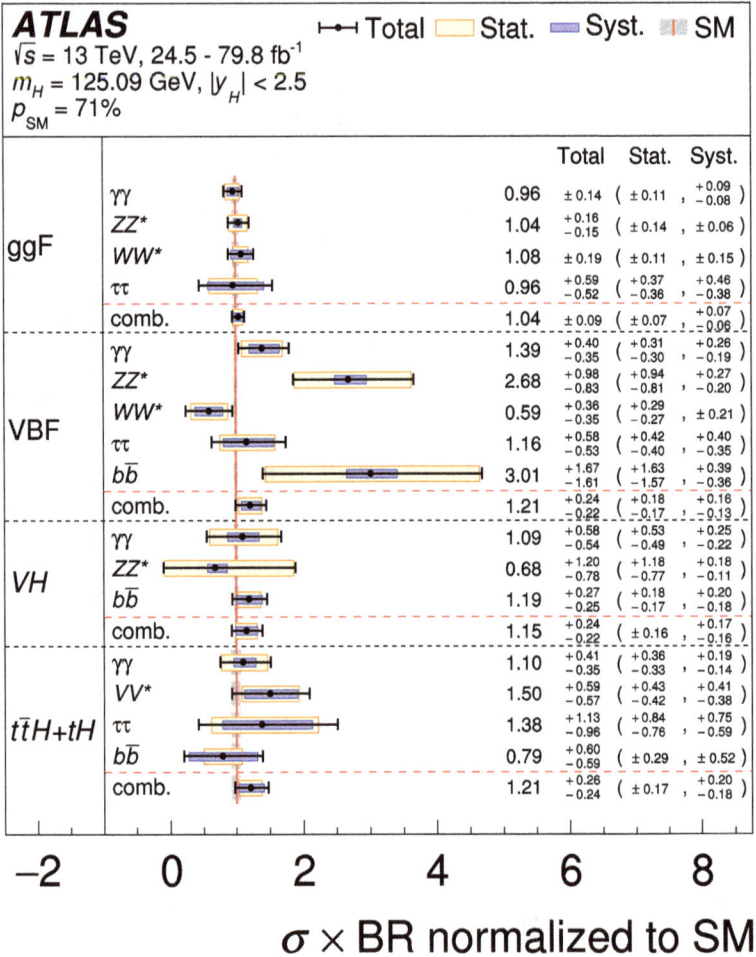

Figure 7.3: The signal strength modifiers μ (cross sections times branching fraction normalized to the standard model prediction) for ggF, VBF, Vh and $t\bar{t}h + th$ production in each relevant decay mode. The values are obtained from a simultaneous fit to all channels. Combined results for each production mode are also shown, assuming standard model values for the branching fractions into each decay mode. The black error bars, blue boxes and yellow boxes show the total, systematic and statistical uncertainties in the measurements, respectively. The gray bands show the theory uncertainties in the predictions.

7.2 The Kappa Framework

The rate of each particular process depends on a number of coupling strengths at production and decay level. To quantify the level of agreement of the experimentally measured couplings with their standard model predicted values a framework of coupling modifiers is introduced. We describe here a framework where only the strength of the couplings is modified, while the tensor structure of the couplings is assumed to be the same as in the standard model prediction. By analogy with the electromagnetic interactions, it would mean that the electric charge is modified, but the functional form of the electric current is not changed. This means in particular that the observed particle (hypothesized to be the Higgs boson) is assumed to be a CP-even scalar, as the standard model predicts. Similarly to the signal strength parameter μ, a coupling modifier is defined as the ratio of the observed value of the Higgs boson coupling with a particle i, g_i^{observed}, to the predicted one, $g_i^{\text{predicted}}$:

$$\kappa_i = \frac{g_i^{\text{observed}}}{g_i^{\text{predicted}}} . \tag{7.2}$$

It must be understood that unlike the signal strength μ parameters, the coupling modifiers κ are not measured directly; and some "unfolding" procedure must be employed. In that sense, these *kappa* modifiers are pseudo-observables. If a large discrepancy from standard model behavior were to be established, this framework would still be useful for assessing the level of compatibility of the model predictions with the data. But to interpret the physical origin of such a discrepancy, one would need to use something other than this framework.

This definition of coupling modifiers, referred to as the kappa-framework, has a rather transparent relation to the observable signal strength if the Higgs boson coupling to particle i happens at tree-level. Let us take as an example $t\bar{t}h$ production with a subsequent Higgs boson decay to a pair of τ leptons. The production process relies on the Higgs boson coupling to the top quark, κ_t, while the decay depends on Higgs boson coupling to a τ lepton, κ_τ. Thus, the observed cross section times branching ratio for this process is related to the predicted one through the following relation:

$$(\sigma_{t\bar{t}h}\text{Br}_{h\to\tau^+\tau^-})^{\text{observed}} = \kappa_t^2\kappa_\tau^2(\sigma_{t\bar{t}h}\text{Br}_{h\to\tau^+\tau^-})^{\text{predicted}}. \qquad (7.3)$$

7.2.1 *Effective loops*

If the Higgs boson does not couple to particle i at the tree-level, such a relationship is less transparent and additional assumptions are necessary. Let us take, for example, Higgs boson production via gluon fusion. In the standard model such a process proceeds via a loop diagram that is dominated by the contribution from the top and bottom quarks (contributions from the second- and in particular first-generation quarks are suppressed due to the small mass of these quarks). If we assume that these couplings are modified with modifiers κ_t and κ_b, respectively, the cross section of Higgs boson production scales as

$$\sigma_{gg\to h}^{\text{observed}} = \sigma^{tt}\kappa_t^2 + \sigma^{bb}\kappa_b^2 + \sigma^{tb}\kappa_t\kappa_b, \qquad (7.4)$$

which includes an interference term σ^{tb}. So we can define an *effective* coupling of the Higgs boson to a gluon κ_g that is a function of couplings κ_t, κ_b and Higgs boson mass, such that:

$$\kappa_g^2(\kappa_t, \kappa_b, m_h) = \frac{\sigma^{tt}\kappa_t^2 + \sigma^{bb}\kappa_b^2 + \sigma^{tb}\kappa_t\kappa_b}{\sigma_{gg\to h}^{\text{predicted}}}. \qquad (7.5)$$

At the same time, one might conceive a theory where the Higgs boson couples to the gluon at tree-level, thus adding an extra contribution to Higgs production via this channel. Thus, a more general modifier κ_g to the Higgs gluon coupling is needed as follows:

$$\kappa_g^2 = \frac{\sigma_{gg\to h}^{\text{observed}}}{\sigma_{gg\to h}^{\text{predicted}}}. \qquad (7.6)$$

Such treatment of the coupling modifier is called the "effective loop" formalism. Similar treatments are applied to $h\gamma\gamma$ and $h\gamma Z$ vertices.

To allow for contributions to Higgs boson decay from particles not predicted by the standard model, the modifier to the total width of the Higgs boson, κ_h^2, is introduced. It is not tied to the other coupling modifiers κ_i, but is rather kept as a free parameter in the global analysis.

7.2.2 *Constraints on κ parameters*

In Tables 7.1 and 7.2, we summarize the most accessible Higgs boson production and decay processes, respectively, and the dependence of their rates on the κ parameters. Even only taking standard model particles into account, we have six κ parameters for quarks, three κ parameters for charged leptons, two κ parameters for W and Z bosons and one for the Higgs boson self-coupling, adding up to 12 free parameters. If we add three more for the effective loops, we have 15 free parameters. Four production mechanisms, gluon fusion, VBF, $t\bar{t}h$ and associated production, depend on a combination of couplings each, as do the decay channels, to vector bosons, and to $b\bar{b}$, $\gamma\gamma$, $\tau\tau$ and γZ. Thus, to extract the information about these couplings, a global fit to the observed vs. predicted rates must be performed.

Some of these decay channels will not be accessible to experimental measurements for quite some time. First-generation fermions have very small standard model expected couplings due to their small mass. Second generation, in particular charm quarks, may be probed in the next 5–10 years. Recently CMS reported

Table 7.1: Dependence of production rates on κ parameters.

Production channel	$\dfrac{\sigma^{\text{observed}}}{\sigma^{\text{predicted}}}$
$gg \to h$	$\kappa_g^2(\kappa_t, \kappa_b, m_h)$
$qq' \to qq'VV \to qq'h$	κ_V^2
$q\bar{q}' \to V^* \to Vh$	κ_V^2
$gg \to t\bar{t}h$	κ_t^2

Table 7.2: Dependence of decay rates on κ parameters. Fermions include $f = u, d, c, s, t, b, e, \mu, \tau$.

Decay channel	$\dfrac{\mathrm{Br}^{\mathrm{observed}}}{\mathrm{Br}^{\mathrm{predicted}}}$
$h \to \gamma\gamma$	$\kappa_\gamma^2(\kappa_t, \kappa_b, \kappa_W, m_h)$
$h \to \gamma Z$	$\kappa_{\gamma Z}^2(\kappa_t, \kappa_b, \kappa_W, m_h)$
$h \to VV$	κ_V^2
$h \to f\bar{f}$	κ_f^2
$h \to hh$	κ_λ^2
Total width	κ_h^2

evidence in the $h \to \mu\mu$ channel based on the full Run 2 statistics. Higgs boson self-coupling is expected to be accessible based on the High-Luminosity LHC data. Meanwhile, some assumptions about the relationship between couplings must be made to extract useful information from the available data. Naturally, once more data are available, more and more assumptions can be relaxed.

It is essential that in the developed theory of electroweak symmetry breaking, the Higgs boson couples to W and Z bosons in the proportion predicted by the model. This is referred to as the custodial symmetry. Thus, inspecting the ratio between κ_W and κ_Z is one of the first checks of the standard model. We can assume a common scaling parameter for all vector bosons κ_V and a universal coupling scaling to fermions κ_f. This assumption is of particular interest since it probes the fermiophobic Higgs models, where a Higgs boson couples only to vector bosons, but not to fermions. We can also decouple the strength of the interactions between quarks and leptons, introducing κ_q and κ_ℓ, respectively. The next step in relaxing the assumptions is to introduce different scaling to up and down-type quarks: κ_{up} and κ_{down}. With each assumption made less and less specific, the ability to test the validity of the standard model increases, but each step also requires additional data to make meaningful statements. In the following, we discuss examples of experimental strategies employed to isolate different production and decay channels, thus allowing access to a particular combination of couplings.

7.3 $h \to WW$

Unlike the $h \to ZZ$ channel, where leptonic final state can be fully reconstructed, establishing a signal in $h \to WW$ channel is harder to achieve. In hadron colliders, since the reacting partons are the constituents of the colliding particles, the total momentum of the reaction in the direction parallel to the beam axis is not a priori known. On the contrary, in the plane transverse to the beam direction, the total momentum of the reaction must be zero. An undetected particle, e.g., neutrino, would result in the transverse momentum imbalance. A characteristic signature of $h \to WW \to \ell^+\nu\ell^-\bar{\nu}$ production is two oppositely charged leptons and missing transverse momentum due to two escaping neutrinos. Such a signature is not unique to Higgs boson production. To distinguish it from background events it is necessary to use kinematic distributions of the final state particles. Leptonic W boson decays contain a charged lepton and a neutrino, which escapes the detector without leaving any signal. Hence, there are three unknown variables — the three components of neutrino momentum. Thus, if both W bosons decay leptonically, there are two neutrinos in the final state, resulting in six unknown parameters. There are four constraints on the system: two requirements that a lepton and a neutrino form a system with an invariant mass of W boson, and the fact that x and y components of the neutrinos add up to balance the corresponding components of the total momentum of the observed particles. Hence, with six unknowns and four constraints, the system is under-constrained making full kinematic reconstruction impossible. However, the so-called transverse mass can be constructed without the knowledge of the neutrinos' momentum parallel to the beam direction:

$$m_{\mathrm{T}} = \sqrt{(E_{\mathrm{T}}^{\ell\ell} + p_{\mathrm{T}}^{\nu\nu})^2 - (\vec{p}_{\mathrm{T}}^{\,\ell\ell} + \vec{p}_{\mathrm{T}}^{\,\nu\nu})^2}, \tag{7.7}$$

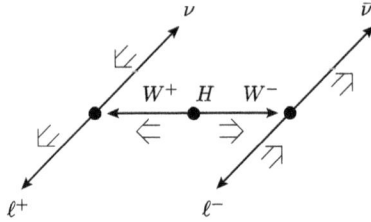

Figure 7.4: Correlation of spins in $h \to WW$ process. The small arrows indicate the particles' directions of motion and the large double arrows indicate their spin projections. The spin-zero Higgs boson decays to W bosons with opposite spins, and the spin-one W bosons decay into leptons with aligned spins. The h and W decays are shown in the decaying particle's rest frame. Because of the $V - A$ decay of the W bosons, the charged leptons have a small opening angle in the laboratory frame.

where $E_{\mathrm{T}}^{\ell\ell} = [(p_{\mathrm{T}}^{\ell\ell})^2 + m_{\ell\ell}^2]^{1/2}$, $m_{\ell\ell}$ is the invariant mass of the two charged leptons, $\vec{p}_{\mathrm{T}}^{\,\nu\nu}$ ($\vec{p}_{\mathrm{T}}^{\,\ell\ell}$) is the vector sum of the neutrino (lepton) momenta and $p_{\mathrm{T}}^{\nu\nu}$ ($p_{\mathrm{T}}^{\ell\ell}$) is its modulus. Unlike the distribution in the invariant mass of the two Z bosons in the case of $h \to ZZ$, which exhibits a narrow peak at the mass of the Higgs boson, the distribution in m_{T} is broad, yet there is a kinematic upper bound at the Higgs boson mass, which is used as a discriminating feature.

The spin-zero Higgs boson decays to W bosons with oppositely pointing spins, and the spin-one W bosons decay into leptons with aligned spins. The h and W decays are shown in the decaying particle's rest frame in Fig. 7.4. The charged leptons have a small opening angle in the laboratory frame due to the $V - A$ nature of the W boson decay.

Ultimately, the information about the lepton opening angles and the missing transverse momentum is used to enhance the signal. The number of events due to Higgs boson production is extracted from the fit of the transverse mass distribution, shown in Fig. 7.5 (left).

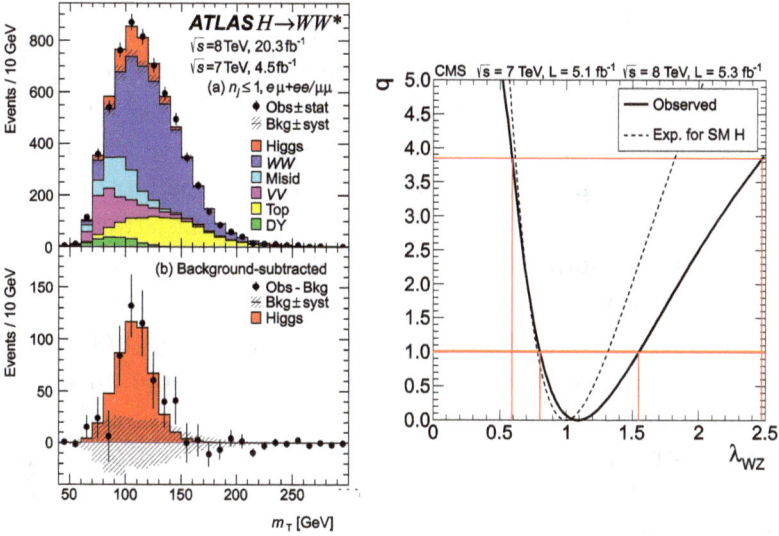

Figure 7.5: (Left) Post-fit combined transverse mass distributions for $N_{\text{jets}} \leq 1$ for all WW lepton-flavor samples in 7 and 8 TeV data from ATLAS. The plot in (b) shows the residuals of the data with respect to the estimated background compared to the expected distribution for an standard model Higgs boson with $m_h = 125$ GeV. (Right) The test statistic $q(\lambda_{WZ})$ as a function of the ratio of the couplings to W and Z bosons, λ_{WZ}, for all $h \rightarrow WW \rightarrow \ell\nu\ell\nu$ and $h \rightarrow ZZ \rightarrow 4\ell$ channels in CMS. The intersection of the curves with the horizontal lines $q = 1$ and 3.8 give the 68% and 95% CL intervals, respectively.

Establishing the $h \rightarrow WW$ signal was of great importance because it allowed to check one of the strongest predictions of the electroweak symmetry breaking theory — the relative coupling between Higgs and W and Higgs and Z bosons. Similar to Eq. (3.14), the standard model predicts the ratio of Higgs boson coupling to W boson to that to Z boson to be equal to $\cos\theta_W$. The ratio of the observed to predicted coupling is parametrized by a coupling modifier, λ_{WZ}, which is expected to be one in the standard model. Thus, even before the evidence for $h \rightarrow WW$ production could be claimed, the relative coupling λ_{WZ} was constrained to be $\lambda_{WZ} = 1.1^{+0.5}_{-0.3}$, as shown in Fig. 7.5 (right).

7.4 The Yukawa Couplings of Fermions

The decay channels of a Higgs boson to a pair of Z or W bosons probe the tree-level couplings to gauge bosons (κ_V). The same is true for the VBF production mechanism. Higgs boson production via gluon fusion and Higgs boson decay to a pair of photons are parametrized by κ_g and κ_γ, respectively. In the standard model, these coupling modifiers depend on the strength of the Higgs boson interaction with the fermions, the Yukawa couplings, the values of which can be extracted from the corresponding production and decay rates. This is demonstrated in Fig. 7.6, where universal coupling modifiers for Higgs boson coupling to fermions κ_f and to vector bosons κ_V are extracted from the combination of the production and decay processes mentioned above. From this analysis it is clear that the fermiophobic Higgs boson scenario, represented in this plot by the magenta dot at $(\kappa_V, \kappa_F) = (1, 0)$, is excluded at 95% CL.

Yet, it is important to establish the Higgs boson coupling to fermions relying on the tree-level processes. Higgs boson decay to a bottom quark pair has the largest branching ratio, followed by a τ lepton pair, both being fermions of the third generation. The identification of b-jets (jets containing b quark hadronization products) and τ leptons is discussed in Boxes 7.1 and 7.2, respectively. Despite large branching fractions, the background processes for these are numerous and to achieve the desired significance, the signals from several production mechanisms were combined — gluon fusion, VBF and associated production. Figure 7.7 shows the distributions in the invariant mass of the two b-jets and the measured μ parameters by production channel. Figure 7.8 shows the invariant mass distributions of two τ leptons and the p-value for Higgs boson decay to a pair of τ leptons.

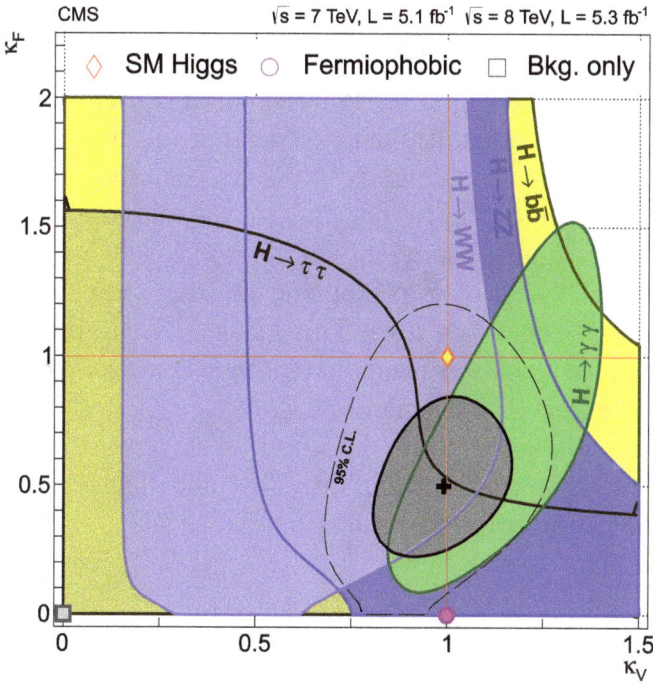

Figure 7.6: The 68% CL contours for the test statistic in the (κ_V, κ_F) plane for individual decay channels (colored regions) and the overall combination (solid thick lines). The thin dashed lines show the 95% CL range for the overall combination. The black cross indicates the global best-fit values. The diamond shows the standard model expectation $(\kappa_V, \kappa_F) = (1, 1)$. The point $(\kappa_V, \kappa_F) = (1, 0)$, indicated by the circle, corresponds to the fermiophobic Higgs boson scenario. It is assumed here that both κ_V and κ_F are positive.

Box 7.1 Identification of *b*-jets

Jets are produced in great numbers in hadronic collisions, most of them are the result of hadronization of light quarks. Bottom quarks have several characteristic features that when combined can provide a reliable discrimination of *b*-jets (jets resulting

(*Continued*)

(*Continued*)

from bottom quark hadronization) from other jets. First, the mass of b quark is significantly larger than lighter quarks. This results in a larger multiplicity of particles in its decays, having more energetic secondary particles and the resultant jet having a larger width. Second, b-hadrons take the majority of the primary quark energy in the hadronization process. And third, probably its most prominent feature is the long lifetime of b hadrons ($c\tau = 454$ μm). This leads to decay products being displaced from the primary vertex, where the b quark was produced. This displacement can be easily measured by modern silicon detectors. This information is typically combined into a multivariate discriminant. The plot illustrates the performance of several b identification algorithms.

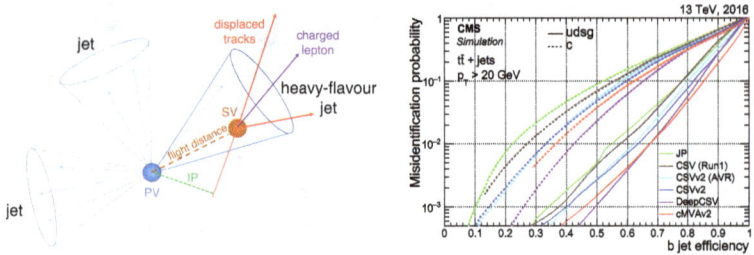

Box 7.2 Identification of τ Leptons

Unlike electrons and muons, τ leptons are not easily identifiable. They are fairly short lived ($c\tau = 87$ μm) and decay either leptonically — to an electron (muon), electron (muon) antineutrino and τ neutrino, or hadronically. Hadronic decays result in a narrow jet, since being a color singlet τ lepton does not produce any products of hadronization process. Typically, there could be one or three charged particles in the final state, referred to as one-prong or three-prong decays, respectively.

(*Continued*)

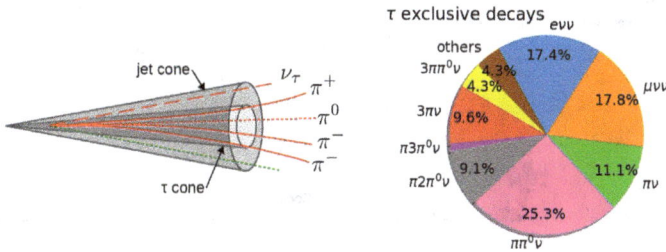

The plot shows an example of a three-prong hadronic τ decay with a narrow τ cone (at the LHC, 90% of the energy is contained in a cone of radius $R = 0.2$ for $p_T^\tau > 50$ GeV) and a wider jet cone which may contain additional EM clusters. τ leptons decay hadronically 65% of the time, and into one charged particle (85%) or three charged particles (15%).

The kinematics of these decay products, the width of the resulting jet and the τ lepton lifetime are used to construct a multivariate discriminant that allows to separate τ leptons from numerous hadronic jets.

7.5 $t\bar{t}h$ Production

Higgs boson production in association with top quarks is the direct proof of its coupling to the top quark. Top quark decays to a b quark and a W boson, which can decay either hadronically or leptonically. Thus, depending on the W boson decay, the signature of top quark pair production is two b-jets accompanied by two leptons, or one lepton and two jets, or four jets. If the final state contains one or two neutrinos, it results in the misbalance of the total momentum in the plane transverse to the beam direction. The top quark was observed via its pair production by the Tevatron experiments CDF and D0 in 1995. Since then its properties have been measured with high precision. In particular, the precise determination of the top

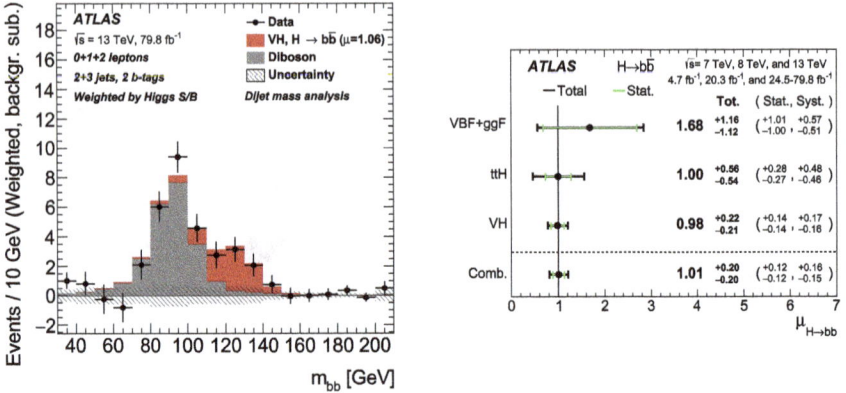

Figure 7.7: (Left) The bb dijet invariant mass distribution in data after subtraction of all backgrounds except for the WZ and ZZ diboson processes, in the $h \to bb$ analysis in ATLAS. The expected contribution of the associated Wh and Zh production of a standard model Higgs boson with $m_h = 125$ GeV is shown scaled by the measured signal strength ($\mu = 1.06$). The size of the combined statistical and systematic uncertainty for the fitted background is indicated by the hatched band. (Right) The fitted values of the Higgs boson signal strength $\mu_{h \to bb}$ for $m_h = 125$ GeV separately for the Vh, $t\bar{t}h$ and VBF+ggF analyses along with their combination, using the 7, 8 and 13 TeV data. The individual $\mu_{h \to bb}$ values for the different production modes are obtained from a simultaneous fit with the signal strengths for each of the processes floating independently. The probability of compatibility of the individual signal strengths is 83%.

quark mass, together with the W boson mass, limited the allowed region for the mass of the Higgs boson, and thus, facilitated the optimization of the search (Section 5.1). The LHC has become a true top quark factory with millions of top pair events observed. Since the Higgs boson coupling to the top quark is so large (of the order of 1 in the standard model), there exists a probability that a Higgs boson is radiated off from a top quark in $t\bar{t}$ events. Yet, due to the high mass of the Higgs boson, these events are very rare. A combination of several Higgs boson decay channels ($b\bar{b}$, WW^*,

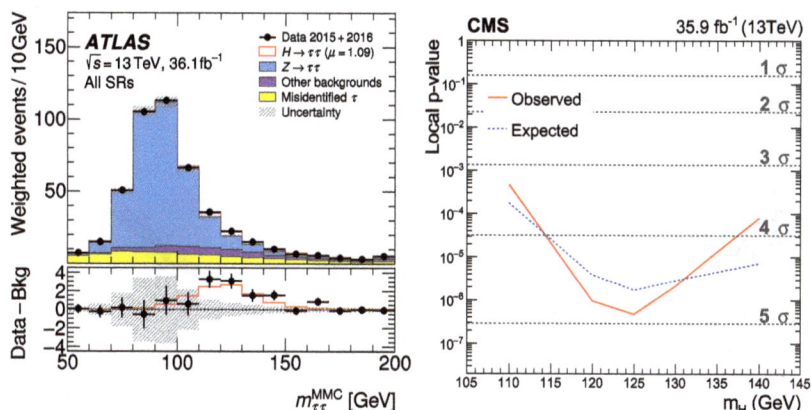

Figure 7.8: (Left) Distribution of the reconstructed $\tau\tau$ invariant mass in ATLAS. The contributions of the different signal regions are weighted by a factor of $\ln(1 + S/B)$, where S and B are the expected numbers of signal and background events in that region, respectively. The bottom panel shows the differences between observed data events and expected background events after applying the same weights (black points). The observed Higgs boson signal ($\mu = 1.09$) is shown with the solid red line. The signal and background predictions are determined in the likelihood fit. The size of the combined statistical, experimental and theoretical uncertainties in the background is indicated by the hatched bands. (Right) Local p-value and significance as a function of the standard model Higgs boson mass hypothesis. The observation (red, solid) is compared to the expectation (blue, dashed) for a Higgs boson with a mass $m_h = 125.09$ GeV. The background includes Higgs boson decays to pairs of W bosons, with $m_h = 125.09$ GeV.

$\tau^+\tau^-$, $\gamma\gamma$ and ZZ^*) was necessary to claim an observation of such events. Despite its low branching ratio, the $h \to \gamma\gamma$ channel is the most sensitive one due to low background and the narrow peak in $\gamma\gamma$ invariant distribution, shown in Fig. 7.9 (left).

The analysis was done at two different center-of-mass energies and the cross sections albeit measured with large uncertainties are in agreement with the standard model predictions, as shown in Fig. 7.9 (right).

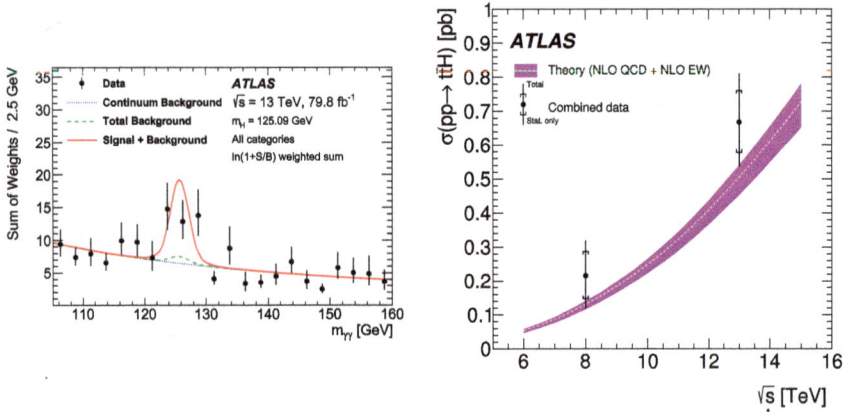

Figure 7.9: (Left) Weighted diphoton invariant mass spectrum in the $t\bar{t}h$-sensitive BDT bins observed in 79.8 fb^{-1} of 13 TeV ATLAS data. Events are weighted by $\ln(1+S_{90}/B_{90})$, where S_{90} (B_{90}) for each BDT bin is the expected $t\bar{t}h$ signal (background) in the smallest $m_{\gamma\gamma}$ window containing 90% of the expected signal. The error bars represent 68% confidence intervals of the weighted sums. The solid red curve shows the fitted signal-plus-background model with $m_h = 125.09 \pm 0.24$ GeV. The non-resonant and total background components of the fit are shown with the dotted blue curve and dashed green curve. Both the signal-plus-background and background-only curves shown here are obtained from the weighted sum of the individual curves in each BDT bin. (Right) Measured $t\bar{t}h$ cross sections in pp collisions at center-of-mass energies of 8 TeV and 13 TeV. Both the total and statistical-only uncertainties are shown. The measurements are compared with the standard model prediction. The band around the prediction represents the PDF+α_S uncertainties and the uncertainties due to missing higher-order corrections.

7.6 Physics Near the $t\bar{t}$ Threshold

The interaction of the Higgs boson with fermions affects not only events where it is "openly" produced and then decays, but also events where particles exchange a virtual Higgs boson. For example, in the production of top quark–antiquark pairs, there is a probability that the top quark emits a virtual Higgs boson and the top antiquark absorbs it, or the other way around (see Fig. 7.10 (left)). The virtual Higgs boson acts as a string between the quark–antiquark pair, slowing them down and pulling them closer to each

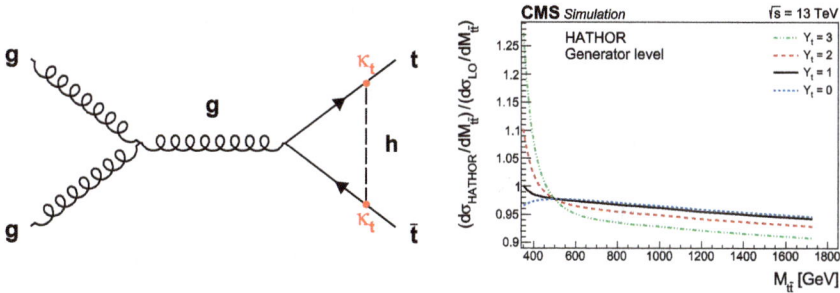

Figure 7.10: (Left) A sample Feynman diagram of $t\bar{t}$ production with a Higgs boson exchange between the top quarks. (Right) The effect of the corrections induced by different strengths of the κ_t coupling (here labeled as Y_t) on the $t\bar{t}$ differential cross section as a function the $t\bar{t}$ invariant mass. A 10% enhancement at parton level (before reconstruction) can occur for $\kappa_t = 2$ near the threshold region (around $2m_t \approx 345$ GeV).

other. In other words, a larger value of the top Yukawa coupling would result in an excess of events with a small invariant mass of the top–antitop quark pair (Fig. 7.10 (right)). CMS searched for the subtle perturbation in the relative motion of top quark and antiquark, and used this information to constrain the ratio of the strength of interaction of the Higgs boson with the top quark to the standard model prediction to be $\kappa_t = 1.16^{+0.24}_{-0.35}$ bounding κ_t to be less than 1.54 at a 95% confidence level. Note that this analysis is truly only sensitive to κ_t, without any assumption about the Higgs coupling to any other particles. The only other channel that can claim this sensitivity only to κ_t is $t\bar{t}h$ production with $h \to t\bar{t}$, also called four-top production in Higgs searches.

In this analysis, the invariant mass of the $t\bar{t}$ system was evaluated using the kinematics of the decay products, which suffers from significant experimental uncertainties, thus limiting the precision of κ_t measurement. This limitation could be avoided if the center-of-mass is known. This is the case in e^+e^- colliders. If such a machine reaches the energy sufficient for $t\bar{t}$ production, it would open possibilities for important precision measurements — top quark Yukawa coupling, top quark mass and width.

7.7 $h \to$ Invisible

Since the Higgs boson generates the mass of all known fundamental particles — gauge bosons and fermions — it is reasonable to assume that it might do so for particles that have not yet been discovered. Thus, the Higgs boson is a portal to the unknown. Of particular interest is dark matter. Astrophysical and cosmological observations suggest that a large component of mass in the universe is invisible, hence the name *dark matter*. If dark matter consists of fundamental particles, such particles do not interact with light (photons). It is also unlikely that these particles interact strongly through gluons, otherwise one would expect them to "clump up" inside hadrons and make a contribution to their mass. Dark matter certainly must interact gravitationally, otherwise it would not be observed in astrophysical effects. A weak interaction of dark matter is not excluded, though not required either. A class of models suggests that dark matter consists of Weakly Interacting Massive Particles (WIMPs). Multiple searches for WIMPs have yielded null results. The Higgs boson discovery, however, opened a new avenue for such searches — the measurement of the Higgs boson invisible width. "Invisible" here implies that the Higgs boson decay products are not registered by detectors, similar to the Z boson invisible width (see Box 2.3), which includes neutrinos and any other particle that escapes detection by conventional means. Since the decay products are not observed, measurements like this are incredibly hard. The Z boson invisible width was measured at an electron–positron collider (LEP), where the total energy and momentum of the collision is known. Still this measurement was a tour de force. At hadron machines such as the LHC, the total energy of the collision and the system's momentum along the beam axis are not known a priori (the proton energy is known, but in each collision the interacting quarks or gluons carry a different fraction of the proton's energy), making such measurement next to impossible.

To be able to identify potential events containing Higgs bosons decaying to invisible particles, it is necessary to have a recoiling system, as is the case for Higgs boson production via VBF, or in association with a vector boson, preferably Z, to avoid additional loss from a neutrino in W boson decay. The recoiling system (two forward jets in the case of VBF or Z boson decay products) is also used to constrain the kinematics of the invisible Higgs boson. The main signature of Higgs boson invisible decay is, therefore, a large missing transverse momentum (see Fig. 7.11). The CMS experiment was able to use these data to constrain the Higgs boson branching ratio to less than 51%. This provides a very stringent limit on dark matter candidates with a mass below half the Higgs boson mass, as shown in Fig. 7.12. Though model-dependent, these

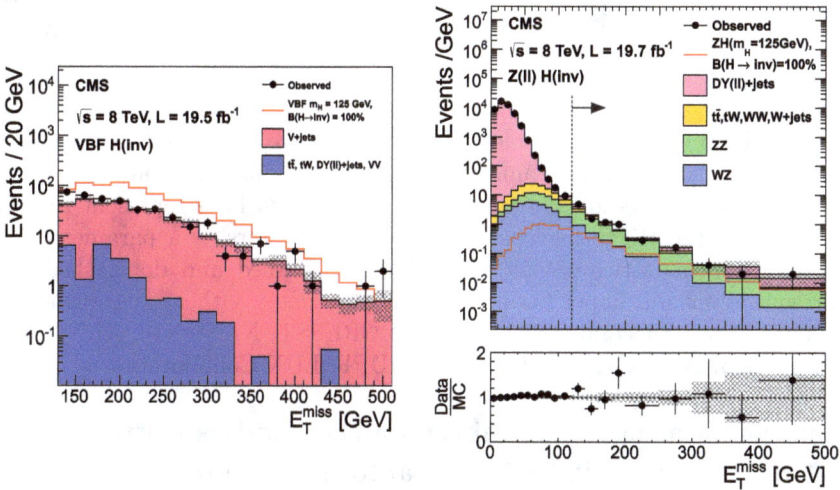

Figure 7.11: The missing momentum (E_T^{miss}) distribution in data and simulation in the VBF $h \to$inv search signal region (left) and the $Z(\to \ell\ell)h(\to$ inv) in CMS (right). The expected distributions from different background processes are displayed cumulatively, while the signal from a Higgs boson with $m_h = 125$ GeV and B($h \to$inv) = 100% is superimposed separately. On the right plot, the arrow corresponds to the cut applied for the final selection.

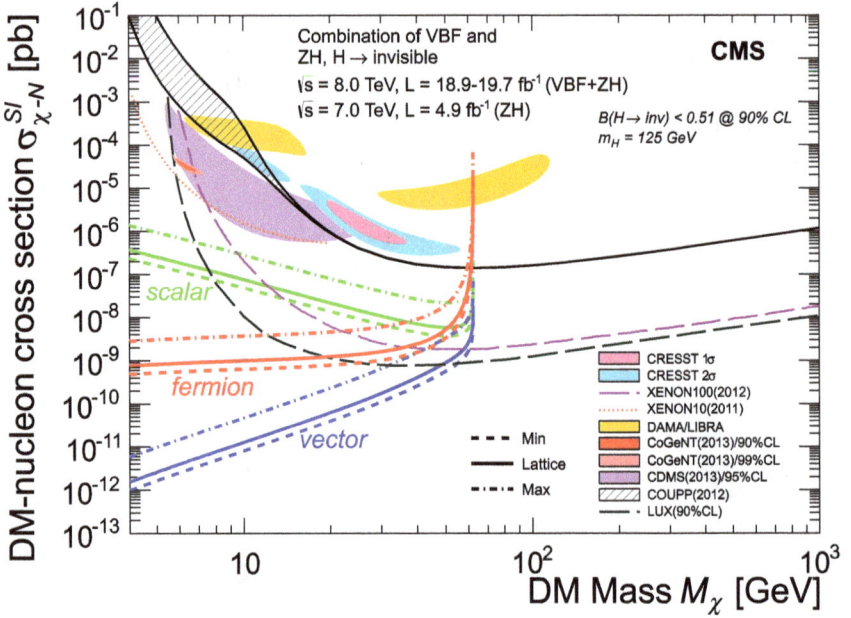

Figure 7.12: Upper limits on the spin-independent dark matter (DM)-nucleon cross section in Higgs-portal models, derived for $m_h = 125$ GeV and B($h \to$ inv)<0.51 at 90% CL, as a function of the DM mass. Limits are shown separately for scalar, vector and fermion DM. The solid lines represent the central value of the Higgs–nucleon coupling, which enters as a parameter, and is taken from a lattice calculation, while the dashed and dot-dashed lines represent lower and upper bounds on this parameter. Other experimental results are shown for comparison, from the CRESST, XENON10, XENON100, DAMA/LIBRA, CoGeNT, CDMS II, COUPP, LUX Collaborations.

limits complement those of direct searches for dark matter, proving that Higgs boson is indeed a portal to the unknown.

7.8 Combined Results

Figure 7.13 summarizes the measured values of the coupling strength vs particle mass for top quark, W and Z bosons, bottom quark, τ and μ leptons. All are in agreement with the standard model predictions.

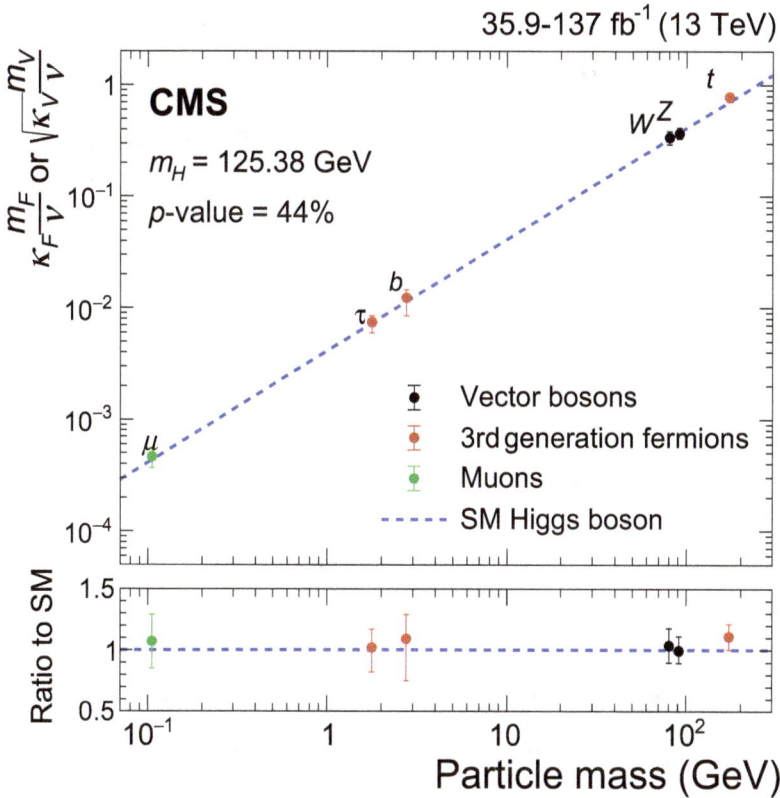

Figure 7.13: The best fit estimates for the reduced coupling modifiers extracted for fermions and weak bosons from the resolved κ-framework compared to their corresponding prediction from the standard model. The error bars represent 68% CL intervals for the measured parameters. In the lower panel, the ratios of the measured coupling modifier values to their SM predictions are shown.

Using the κ framework, a combined measurement of couplings was performed in all different production processes and decay modes. Assuming common coupling modifiers for all fermions, κ_F, and for all electroweak gauge bosons, κ_V, results in the allowed region shown in Fig. 7.14. The black cross marks the standard model value. The measured values are in a very good agreement with this prediction.

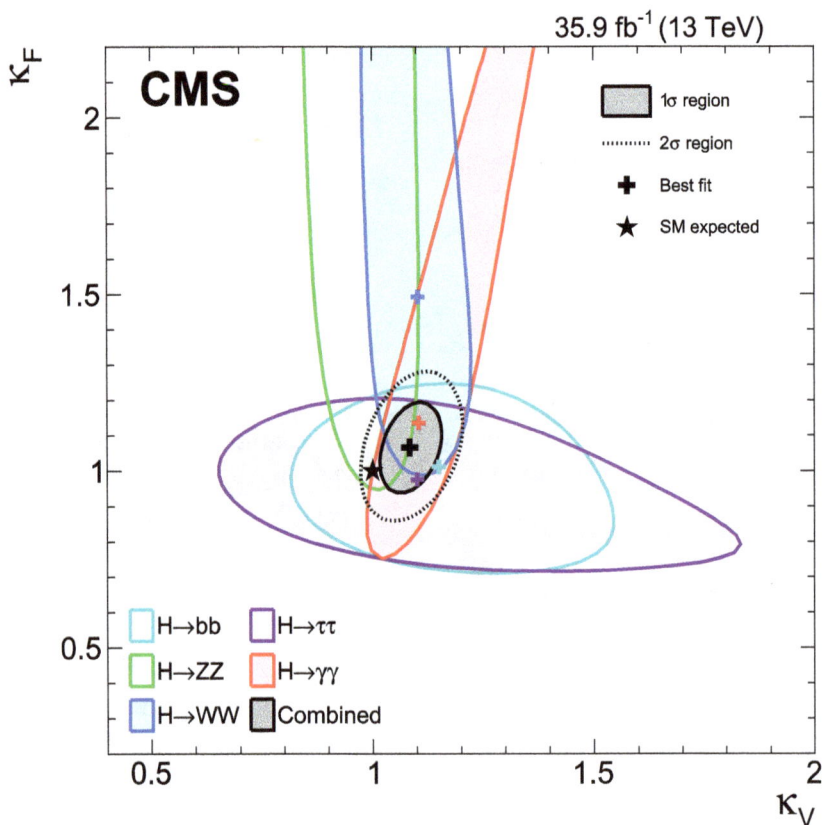

Figure 7.14: The 1σ and 2σ CL regions in the κ_F vs. κ_V parameter space for a model assuming a common scaling of all the vector boson and fermion couplings. Note the improvement with respect to Fig. 7.6 due to new data and the combination of improved analyses.

Suggested Reading for Chapter 7

[1] ATLAS Collaboration. "Measurement of the Higgs boson coupling properties in the $H \rightarrow ZZ^* \rightarrow 4\ell$ decay channel at $\sqrt{s} = 13\,\text{TeV}$ with the ATLAS detector". *JHEP* 03 (2018), p. 095. arXiv: 1712.02304 [hep-ex].

[2] ATLAS Collaboration. "Combined measurements of Higgs boson production and decay using up to $80\,\text{fb}^{-1}$ of proton-proton

collision data at $\sqrt{s} = 13\,\mathrm{TeV}$ collected with the ATLAS experiment". *Phys. Rev. D* 101.1 (2020), p. 012002. arXiv: 1909.02845 [hep-ex].

[3] CMS Collaboration. "Combined measurements of Higgs boson couplings in proton-proton collisions at $\sqrt{s} = 13\,\mathrm{TeV}$". *Eur. Phys. J. C* 79.5 (2019), p. 421. arXiv: 1809.10733 [hep-ex].

[4] ATLAS Collaboration. "Observation and measurement of Higgs boson decays to WW* with the ATLAS detector". *Phys. Rev. D* 92.1 (2015), p. 012006. arXiv: 1412.2641 [hep-ex].

[5] ATLAS Collaboration. "Measurements of Higgs boson production and couplings in diboson final states with the ATLAS detector at the LHC". *Phys. Lett. B* 726 (2013). [Erratum: Phys.Lett.B 734, 406–406 (2014)], pp. 88–119. arXiv: 1307.1427 [hep-ex].

[6] ATLAS Collaboration. "Observation of $H \to b\bar{b}$ decays and VH production with the ATLAS detector". *Phys. Lett. B* 786 (2018), pp. 59–86. arXiv: 1808.08238 [hep-ex].

[7] ATLAS Collaboration. "Cross-section measurements of the Higgs boson decaying into a pair of τ-leptons in proton-proton collisions at $\sqrt{s} = 13\,\mathrm{TeV}$ with the ATLAS detector". *Phys. Rev. D* 99 (2019), p. 072001. arXiv: 1811.08856 [hep-ex].

[8] CMS Collaboration. "Observation of the Higgs boson decay to a pair of τ leptons with the CMS detector". *Phys. Lett. B* 779 (2018), pp. 283–316. arXiv: 1708.00373 [hep-ex].

[9] CMS Collaboration. "Identification of heavy-flavour jets with the CMS detector in pp collisions at $13\,\mathrm{TeV}$". *JINST* 13.05 (2018), P05011. arXiv: 1712.07158 [physics.ins-det].

[10] CMS Collaboration. "Reconstruction and identification of τ lepton decays to hadrons and ν_{tau} at CMS". *JINST* 11.01 (2016), P01019. arXiv: 1510.07488 [physics.ins-det].

[11] ATLAS Collaboration. "Observation of Higgs boson production in association with a top quark pair at the LHC with the ATLAS detector". *Phys. Lett. B* 784 (2018), pp. 173–191. arXiv: 1806.00425 [hep-ex].

[12] CMS Collaboration. "Measurement of the top quark Yukawa coupling from $t\bar{t}$ kinematic distributions in the dilepton final state in proton-proton collisions at $\sqrt{s} = 13\,\text{TeV}$". *Phys. Rev. D* 102.9 (2020), p. 092013. arXiv: 2009.07123 [hep-ex].

[13] Martin Beneke et al. "Higgs effects in top anti-top production near threshold in e^+e^-. annihilation". *Nucl. Phys. B* 899 (2015), pp. 180–193. arXiv: 1506.06865 [hep-ph].

[14] CMS Collaboration. "Search for invisible decays of Higgs bosons in the vector boson fusion and associated ZH production modes". *Eur. Phys. J. C* 74 (2014), p. 2980. arXiv: 1404.1344 [hep-ex].

[15] CMS Collaboration. "Evidence for Higgs boson decay to a pair of muons". *JHEP* 01 (2021), p. 148. arXiv: 2009.04363 [hep-ex].

[16] N. Belyaev, R. Konoplich, and K. Prokofiev. "Study of kinematic observables sentitive to the Higgs boson production channel in $pp \rightarrow Hjj$ process". *J. Phys. Conf. Ser.* 934.1 (2017), p. 012030.

[17] Matthew J. Strassler and Michael E. Peskin. "The Heavy top quark threshold: QCD and the Higgs". *Phys. Rev. D* 43 (1991), pp. 1500–1514.

Chapter 8

Future Measurements

8.1 Accessing Higgs Couplings to the Second-Generation Fermions

It is important to verify that the coupling of the Higgs boson is not limited to the third generation of fermions. Yet, experimentally these measurements are incredibly difficult since the branching ratios to the second (not to mention first)-generation fermions are very small. Background processes are also numerous. For this reason, significant statistics must be accumulated for unambiguous claims of an observation to be made.

8.1.1 $h \to \mu\mu$

Muon identification is discussed in Box 8.1.

Box 8.1 Identification of Muons

Muons are probably the easiest elementary particles to identify. Muons are charged and thus produce tracks in tracking devices. They are fairly long-lived, and thus typically decay outside of the detector volume. Being more massive, they do not radiate as much as electrons. They do not carry any color charge, hence they do not interact strongly with

(Continued)

(*Continued*)

nuclear matter. In fact, aside from elusive neutrinos, muons are the only particles capable of penetrating the material in the calorimeters, magnet and its return yolk. For this reason, tracking devices installed within and outside of the return yolk are referred to as muon chambers.

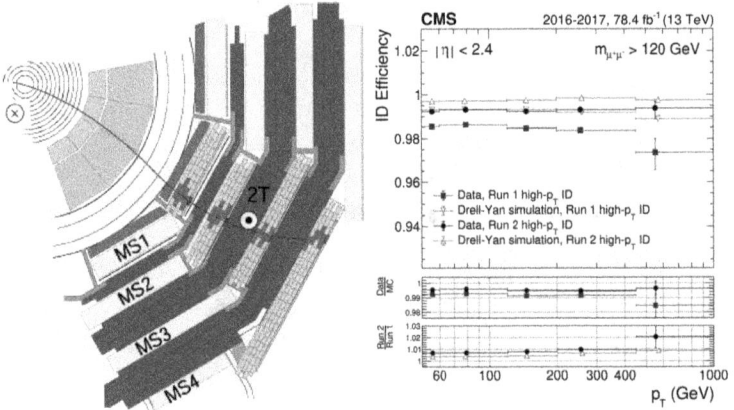

The plot on the left illustrates the basis of muon identification in CMS with drift tubes and cathode strip chambers interspersed with the iron of the return yolk. The magnetic field inside the solenoid and in the return yolk create the distinctive S shaped track for muons. The plot on the right shows the performance of muon identification in the CMS experiment as a function of the transverse momentum of the muon. The performance of different algorithms from Run 1 (2010–2012) and Run 2 (2015–2018) are shown, for simulation and data.

Combining four Higgs boson production mechanisms — gluon fusion, VBF, associated production with vector bosons and production in association with top quark pairs — the CMS collaboration was able to isolate Higgs boson decays to a pair of muons at the

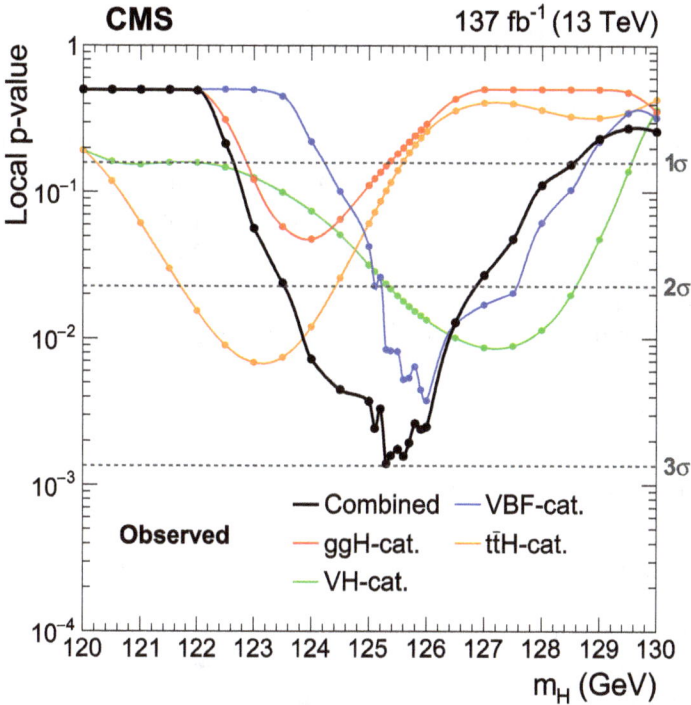

Figure 8.1: Observed local p-values as a function of m_h, extracted from the combined fit as well as from each individual production category in the search for $h \to \mu\mu$, are shown. The solid markers indicate the mass points for which the observed p-values are computed.

3σ level as shown in Fig. 8.1. This means that the probability for the background to fluctuate to mimic this signal is less than 0.2%.

8.1.2 $h \to c\bar{c}$

While experiments can identify first- and second-generation leptons: electrons and muons, with good efficiency and low fake rate, identification of jets resulting from hadronization of charm, strange, up and down quarks is very hard. Out of the listed hadronic modes, the Higgs boson decay to a charm–anticharm quark pair is the dominant one. Charm identification is discussed in Box 8.2.

Box 8.2 Identification of *c*-jets

The charm quark is characterized by a significant lifetime ($c\tau = 120$–300 μm), which is smaller than that of bottom quark, but longer compared to lighter quarks. The charm quark mass is another distinguishing factor — smaller than that of bottom quark and significantly larger than that of strange, up and down quarks. The charm quark fragmentation pattern can also be used to distinguish the corresponding jets. While most of the energy of lighter quarks is lost to the hadronization process, charm hadrons tend to carry a significant portion of the energy of the charm quark. None of these features alone is sufficient to give a reliable discrimination of charm jets from the other jets, yet combined using multivariate techniques, they can provide powerful discrimination.

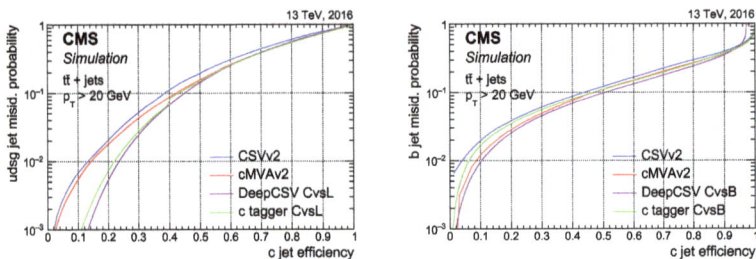

This plot shows the result of training Deep Neural Nets (DeepCSV) to discriminate charm jets vs. light jets (left) and vs. bottom quark jets (right).

As an example, a search for Higgs boson produced in association with a vector boson and decaying into a charm quark pair performed on 35 fb^{-1} of LHC data produced an expected limit that is 37 times larger than the cross section predicted by the standard model (see Fig. 8.2). Clearly, a significant increase in statistics is required to establish this signal.

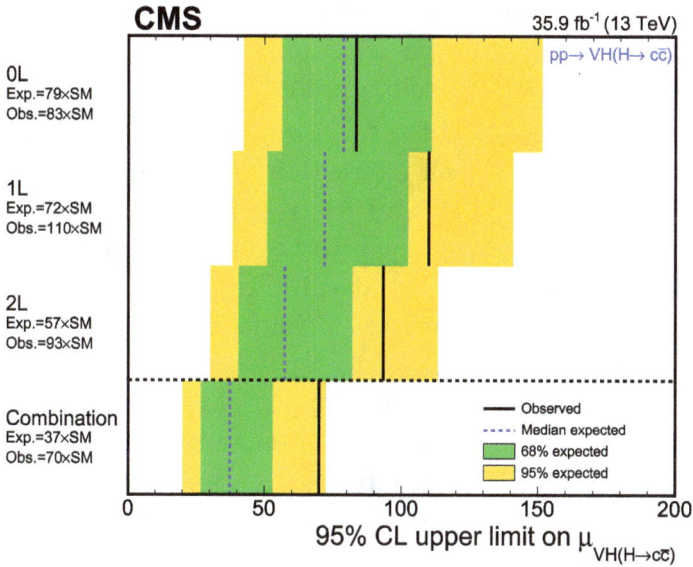

Figure 8.2: The 95% CL upper limits on the μ cross section ratio with the standard model for the $Vh(h \to c\bar{c})$ process from the combination of the resolved-jet and merged-jet topologies analyses in the different final states based on the number of leptons (0L, 1L and 2L depending on how the vector boson $V = W, Z$ decay), and combined. The inner (green) and the outer (yellow) bands indicate the regions containing 68% and 95%, respectively, of the distribution of limits expected under the background-only hypothesis.

8.2 Higgs Self-coupling — Mapping the Higgs Potential

So far we discussed the interactions of the Higgs boson with gauge bosons and fermions. As it turns out, it also interacts with itself. Expanding Eq. (3.2) around the minimum of ϕ, given in Eq. (3.5), we get

$$V(h) = -\frac{\mu^4}{4\lambda} - \mu^2 h^2 + \lambda_3 v h^3 + \frac{\lambda_4}{4} h^4 \,. \tag{8.1}$$

Here, we took into account that $v^2 = -\frac{\mu^2}{\lambda}$. The first term is just a constant. Since $V(\phi)$ can be modified by adding a constant term without changing the equations of motion, the first term can be removed by adding a proper constant to the definition of potential. The second term is reinterpreted as the Higgs boson mass term $\frac{m_h^2}{2}h^2$. The third and the forth terms represent the triple and quadruple Higgs boson self-interactions, for which we introduced the coupling constants λ_3 and λ_4, respectively. In the standard model, λ_3 and λ_4 are both equal to λ. Any deviation of the Higgs potential from the form postulated in Eq. (3.2) would result in λ_3 and λ_4 not being equal. Hence, an independent measurement of these two constants is essential to verify the form of the Higgs potential. The triple Higgs boson coupling is probed via observation of double Higgs boson production, and the quadruple coupling via triple Higgs production. These processes are extremely rare and require significant amount of data. An additional complication comes from the fact that there are other diagrams that interfere with the triple and quadruple Higgs boson coupling, as shown in Fig. 8.3. This means that assumptions must be made about the strength of Higgs boson coupling to other particles.

First attempts to identify double-Higgs production with one Higgs boson decaying into a pair of photons and the other one to

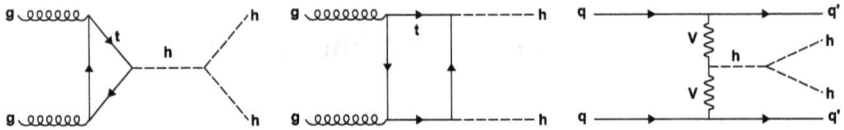

Figure 8.3: Feynman diagrams for di-Higgs production at hadron colliders via the gluon–gluon fusion (left and center) and VBF (right), at leading order. Some processes involve the triple Higgs coupling (left and right), and others involve couplings to other particles (like the top quark Yukawa coupling, in the center plot). Due to the interference between the left and center diagrams, when extracting λ_3, some assumption must be made about κ_t. More complicated diagrams involving a quadratic Higgs coupling can give rise to three Higgs bosons in the final state, for example, by radiating a third Higgs boson in the last vertex of the left diagram.

a pair of bottom quarks constrained κ_λ to be between -3.3 and 8.5 (see Fig. 8.4). It is clear that to get to κ_λ values close to one, significantly more data are needed.

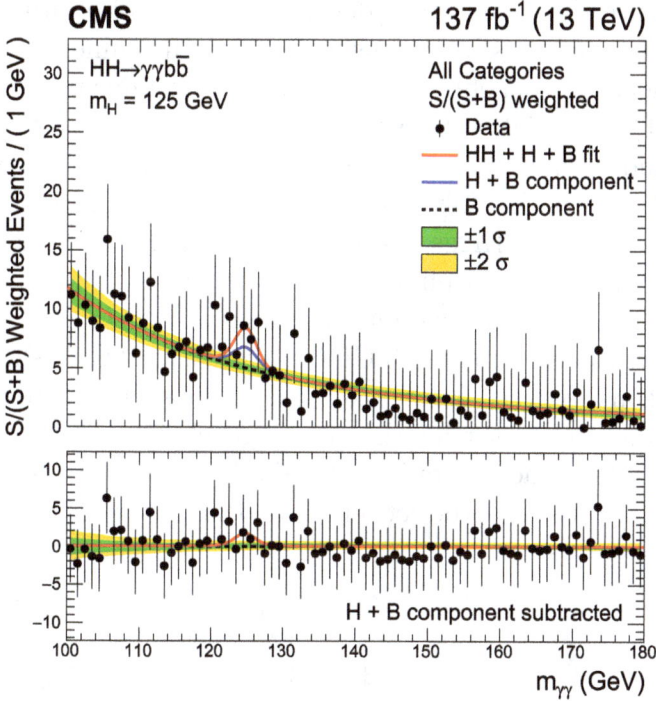

Figure 8.4: Distributions in the invariant mass of two photons, $m_{\gamma\gamma}$ for the selected events in data (black points) weighted by S/(S + B), where S(B) is the number of signal (background) events extracted from the signal-plus-background fit in CMS data. The solid red line shows the sum of the fitted signal and background (HH + H + B), the solid blue line shows the background component from the single Higgs boson and the non-resonant processes (H + B), and the dashed black line shows the non-resonant background component (B). The normalization of each component (HH, H, B) is extracted from the combined fit to the data in all analysis categories. The one (green) and two (yellow) standard deviation bands include the uncertainties in the background component of the fit. The lower panel shows the residual signal yield after the background (H + B) subtraction.

8.3 Considerations for Future Colliders

Increasing the luminosity, as well as center-of-mass energy, in hadronic colliders will lead to an increased observed number of events of a particular process (see Fig. 8.5), and will therefore also improve the precision in determining Higgs boson properties. Yet, studies of Higgs bosons using a lepton machine could deliver complementary benefits. A cleaner environment, well-known energy of the initial state particles, and the fact that their entire energy and momentum are passed on to the final state products, are hard to match. Table 8.1 summarizes the production cross section for various Higgs boson production mechanisms in past, current and future hadron and e^+e^- colliders.

We have seen that a circular machine — LEP II with the maximum energy per beam of 103 GeV — was just 10 GeV short of the Higgs boson discovery. Energy loss due to the synchrotron radiation was the limiting factor for the energy reach of LEP II. A modest increase in the machine radius would allow to reach the threshold for the Higgs boson production in association with the Z boson.

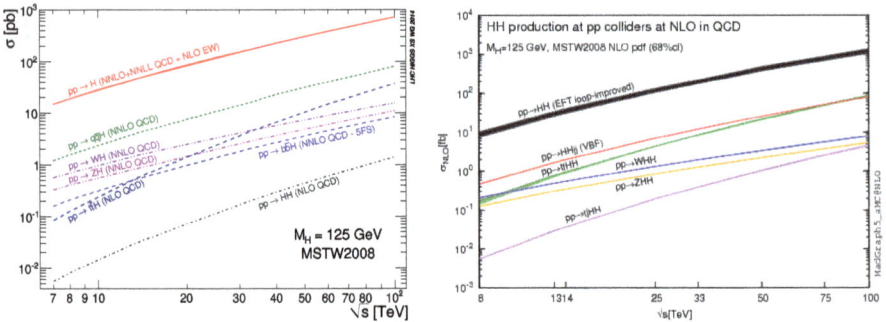

Figure 8.5: (Left) Higgs boson production cross sections as a function of center-of-mass energies ($\sqrt{s}$) at pp colliders. (Right) Total cross sections at next-to-leading order in QCD for the six largest hh production channels at pp colliders for $\sqrt{s}$ from 8 to 100 TeV. The thickness of the lines corresponds to the scale and PDF uncertainties added linearly.

Table 8.1: Higgs production cross sections in pb at various colliders, for $m_h = 125$ GeV. The Future Circular Collider (FCC) is proposed to have up to 100 km radius and reach $\sqrt{s} = 100$ TeV if colliding protons (FCC-hh), or various center-of-mass energies if run as an e^+e^- collider (FCC-ee). Other proposed linear e^+e^- accelerators are the ILC and CLIC. Note that the instantaneous luminosity for an e^+e^- collider could vary from 2×10^{34} up to 2×10^{36} cm^{-2}s^{-1}, and may also include polarized beams (for example, an -80% polarization of the electron beam can increase the $e^+e^- \to hh\nu\nu$ cross section by a factor 1.8, and the $e^+e^- \to hhZ$ cross section by a factor 1.12).

Prod. mode	Tevatron 1.96 TeV	LHC 7 TeV	LHC 14 TeV	FCC-hh 100 TeV	e^+e^-
h	0.950	16.85	54.67	740.3	0.0003 ($\sqrt{s} = 125$ GeV)
VBF	0.067	1.241	4.28	82.0	0.030 ($\sqrt{s} = 350$ GeV)
Wh	0.130	0.577	1.51	15.90	
Zh	0.079	0.339	0.986	11.26	0.200 ($\sqrt{s} = 240$ GeV)
tth	0.004	0.088	0.614	37.9	
hh	—	0.005	0.034	1.224	
$hhqq$	—	—	0.011	0.083	
hhZ			0.001	0.016	0.0002 ($\sqrt{s} = 500$ GeV)
$hh\nu\nu$			—	—	0.0006 ($\sqrt{s} = 3$ TeV)

Such a machine would allow to study the properties of a singly produced Higgs boson in the clean environment of an e^+e^- collider. A precision measurement of the dependence of WW production on the center-of-mass energy and possibly improved precision in the $\sin\theta_W$ adds to the physics program of such a machine. But is it worth the significant expense of a new collider construction? It will not allow to probe the Higgs potential via multiple Higgs boson productions, nor will it reach the $t\bar{t}$ threshold. In e^+e^- colliders, two main production processes can be used to measure the Higgs self-coupling: the Higgsstrahlung-like process $e^+e^- \to Zhh$, which can be probed at $\sqrt{s} \approx 500$ GeV, and double-Higgs boson production in the WW-fusion process $e^+e^- \to hh\nu\nu$, which becomes dominant

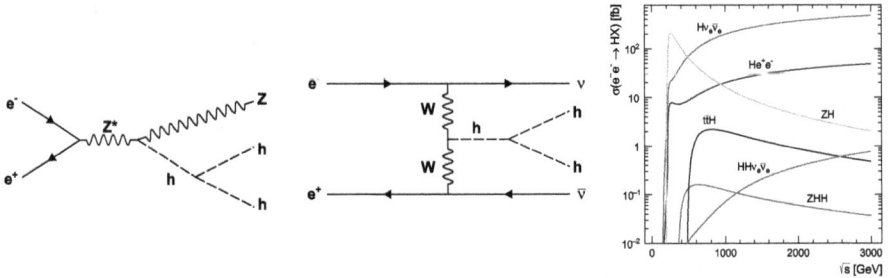

Figure 8.6: (Left and Center) The two main Feynman diagrams for di-Higgs production in e^+e^- collisions, Zhh and $hh\nu\nu$. (Right) Production cross section as a function of center-of-mass energy for the main Higgs production processes at an e^+e^- collider. The values shown correspond to unpolarized beams with initial-state radiation and do not include the effect of beamstrahlung.

above 1.1 TeV. These two processes and their relative production cross sections are shown in Fig. 8.6. To get to such energies with a circular machine is hardly feasible. A linear collider, on the other hand, though technically a lot more challenging, would provide a richer physics program. Scientists around the world are engaged in such a debate.

8.4 Overconstraining the Standard Model

Despite having many free parameters that have to be constrained based on experimental data, the standard model has an incredible predictive power. We have discussed in Section 5.1 that the values of W and Z bosons, top quark and later of Higgs boson masses were predicted with good precision prior to their discoveries. Now, with the discovery of the Higgs boson and an experimental evaluation of its mass, the theory becomes overconstrained. This means we can evaluate if the available measurements are consistent with each other within their uncertainties. If not, it would signal that the standard model needs to be extended.

Without diving too deep into the anatomy of global fits, let us comment on what is the connection between the observables and the model parameters. Radiative corrections to the W boson mass m_W depend on the top quark and Higgs boson masses, m_t and m_h, respectively. The ratio of Z and W boson masses depends on the weak mixing angle, θ_W, as do total decay widths of W, Γ_W and Z bosons, Γ_Z. The asymmetries of fermion f pair production at the Z pole (A_f), and absolute (σ_{hadr}) and relative rates (R_f) measured by the LEP and SLC experiments, also provide powerful constraints on the weak mixing angle. These observables for quarks depend on the quark masses, which, though are not dependent on the electroweak parameters, have to be provided to the fit as nuisance parameters. More information on θ_W can be extracted from the angular distributions of the decay products of W and Z bosons produced in hadronic collisions. The strength of the electromagnetic (α) and strong (α_S) interactions evaluated at the Z pole also add sensitivity. Overall, the χ^2 was found to be equal to 18.6 for 15 degrees of freedom (see Fig. 8.7 (left)). The largest deviation of the fitted value with the input measurement corresponds to the asymmetry in the production of b quarks measured at the Z pole, which amount to 2.4 standard deviations. Given the number of different measurements that were used by the fit, it is not surprising to find at least one such deviation.

Another way to test the internal consistency is to drop one of the inputs from the fit and see how far the prediction for it (indirect measurement) lands from the observed value. This is shown in the right of Fig. 8.7. Take for example the Higgs boson mass. The indirect value for it is 90^{+21}_{-18} GeV, which is in agreement with the direct measurement within 1.7 standard deviations. Overall, the result of the global fit signals a remarkable consistency of the observational data with the standard model.

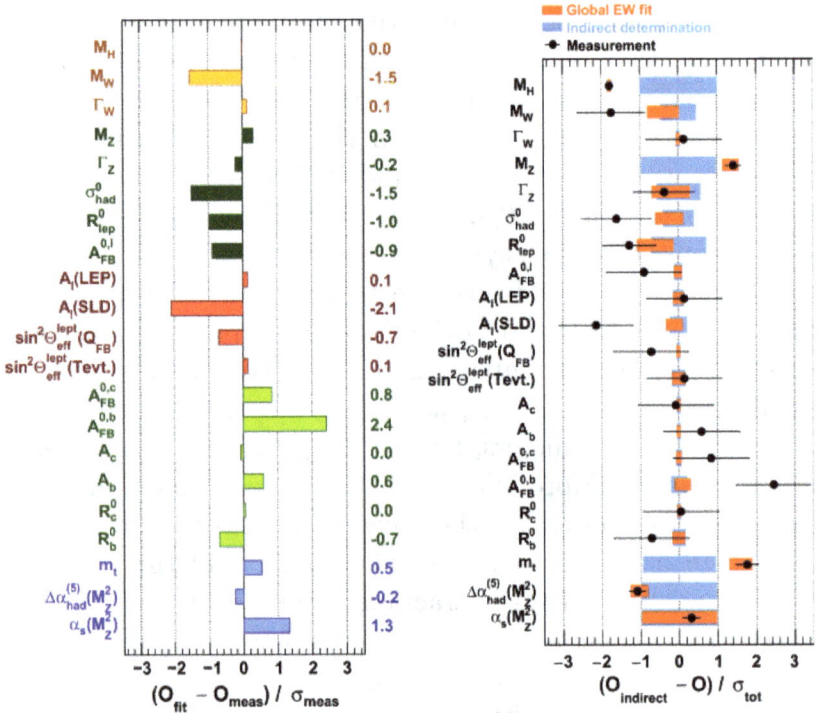

Figure 8.7: Result of the global fit to electroweak parameters. (Left) Difference of the global fit result of an observable and the corresponding input measurement in units of the measurement uncertainty. (Right) The difference between the indirect and direct measurement of the parameters in units of their uncertainty.

Suggested Reading for Chapter 8

[1] CMS Collaboration. "Performance of the reconstruction and identification of high-momentum muons in proton-proton collisions at $\sqrt{s} = 13\,\mathrm{TeV}$". *JINST* 15.02 (2020), P02027. arXiv: 1912.03516 [physics.ins-det].

[2] CMS Collaboration. "A search for the standard model Higgs boson decaying to charm quarks". *JHEP* 03 (2020), p. 131. arXiv: 1912.01662 [hep-ex].

[3] CMS Collaboration. "Search for nonresonant Higgs boson pair production in final states with two bottom quarks and two photons in proton-proton collisions at $\sqrt{s} = 13\,\text{TeV}$". *JHEP* 03 (2021), p. 257. arXiv: 2011.12373 [hep-ex].

[4] R. Frederix et al. "Higgs pair production at the LHC with NLO and parton-shower effects". *Phys. Lett. B* 732 (2014), pp. 142–149. arXiv: 1401.7340 [hep-ph].

[5] CLICdp, CLIC Collaboration. "The Compact Linear Collider (CLIC) — 2018 Summary Report". *CERN Yellow Report* 2/2018 (Dec. 2018), arXiv: 1812.06018 [physics.acc-ph].

[6] Johannes Haller et al. "Update of the global electroweak fit and constraints on two-Higgs-doublet models". *Eur. Phys. J. C* 78.8 (2018), p. 675. arXiv: 1803.01853 [hep-ph].

Chapter 9

Scalar Fields Beyond the Standard Model

9.1 Why the Standard Model is not Complete

Within the standard model, we have seen how elegantly the Higgs mechanism describes the way in which particles acquire their mass by introducing a new scalar field that breaks the electroweak symmetry, leaving the Lagrangian invariant under gauge transformations. The standard model has been tested to an incredible level of precision for many decades and so far its predictions seem to agree with the experiments. And yet, despite its amazing success, there are some lingering issues that make it impossible to assume that this theory is a complete description of Nature. For example, it does not explain the different values of the quantum numbers like the electric charge Q, weak isospin I or hypercharge Y that each particle has. In addition, it contains at least 19 arbitrary parameters: the three independent gauge couplings, six quark and three charged lepton masses, three Cabibbo weak mixing angles, one CP-violating quark mixing phase, one possible CP-violating parameter in the strong interaction, and finally, two independent masses for the weak bosons.

In the standard model, the masses of fermions are generated via a Yukawa coupling of the scalar Higgs doublet Φ with the right-handed and left-handed components of the fermion. The right-handed component is an $SU(2)_L$ singlet, the left-handed component is part of a doublet. As introduced in Eq. (3.16) for leptons, we can write:

$$\mathcal{L}_{\text{Yukawa}} = -g_e \left[\bar{e}_L \phi e_R + \bar{e}_R \bar{\phi} e_L \right], \tag{9.1}$$

where e_L is the left-handed lepton doublet and e_R is the right-handed charged lepton field, and there are three generations of each field. After spontaneous symmetry breaking these terms lead to charged lepton masses of the form $m_\ell = g_\ell v/\sqrt{2}$, proportional to the vev. Since the standard model only contains left-handed neutrinos (there is no right-handed neutrino singlet ν_R), neutrinos remain massless in the theory. It is impossible to add a mass term of the form $m\bar{\psi}_R \psi_L$ without a suitable right-handed partner for the neutrino. We know, however, that neutrino flavor eigenstates mix to form neutrino mass eigenstates (like quarks mix in the CKM matrix), because we have observed their oscillation from one flavor to another, which is only possible if they have mass. Therefore, some mechanism, beyond the standard model, must be responsible for neutrino masses.

From precise measurements of tritium β-decays, the electron-neutrino mass is constrained to be smaller than 1 eV. And similar upper limits have now been obtained on the sum of the three neutrino masses from Cosmological data. Then, why are neutrinos at least five orders of magnitude lighter than electrons? (For comparison, the top and bottom quark masses "only" vary by a factor of 40.) The standard model needs to be extended to include even more parameters: three neutrino masses and three CP-violating phases, but more importantly, a new mechanism for them to acquire their mass. One option would be a "sterile" neutrino, a right-handed SU(2) singlet with Yukawa couplings to the neutral component of the Higgs doublet, but no interactions with other standard model particles. This would create a suitable partner to add a mass term together with the normal "active" (isospin 1/2) left-handed neutrinos. Of course, one then still needs to explain why the Yukawa coupling is that much weaker than the electron's coupling. A second option would be if the neutrino and antineutrino are the same particle (this is called a Majorana fermion). Then it is possible to add a Majorana mass term ($m\chi\chi$, note the two identical fields, not one complex conjugated like for the Dirac spinor $\bar{\psi}$ above) with the left-handed neutrino and a right-handed sterile neutrino that does not violate electroweak symmetry breaking. In this case, the gauge symmetry is preserved and only charge is not conserved (if χ carries a charge Q, like fermion number, the term $m\chi\chi$ "creates" a non-zero charge of 2Q, so charge conservation is violated by this term).

Going beyond particle physics, the standard model does not include a description of gravity, which one would like to be included in a grand unified theory, along with the other forces. At least two more parameters would be needed for this: the coupling strength of gravity given by Newton's constant in terms of the Planck mass $G_N = 1/m_P^2$ (with $m_P \sim 10^{19}$ GeV), and some form of time-varying vacuum energy or cosmological constant Λ to account for the recent cosmological data that support an accelerated expansion of the Universe. All this still leaves out a description of dark matter, which presumably would add even more parameters to the theory.

All these problems of physics beyond the standard model can be grouped into three big categories: (i) Unification, is there a group symmetry that can unify all particles and interactions, a so-called Grand Unified Theory (GUT)?; (ii) Flavor, why are there so many different quarks and leptons and why do they mix so particularly under the weak interaction?; and (iii) Mass: why do particles have such different masses and, in particular, why are the W, Z and Higgs bosons so much lighter than the Planck mass?

We will focus here only on the last item, and how new scalar fields may be introduced to solve some of the problems in the standard model.

9.2 The Hierarchy Problem and Fine Tuning

We have seen that the Higgs potential has a mass parameter m_h^2 expected to be proportional to the square of the vacuum expectation value of the field $m_h^2 \propto -v^2$. This simple picture is fine at leading order, however, problems arise when we account for quantum (loop) corrections. Indeed, the mass term of the Higgs potential receives quantum corrections from the standard model particles, such as W, Z, top quarks and the Higgs boson itself, as depicted in Fig. 9.1.

All three diagrams result in quadratically divergent contributions $m_h^2 + \Delta m_h^2 = m_h^2 + \mathcal{O}(\frac{g}{\pi})\Lambda^2$, dependent on a cutoff energy scale Λ which is the energy up to which the standard model remains valid. It is thus reasonable to require that these quantum corrections should not be too large compared with the original mass, which means $\Lambda \lesssim 1$ TeV for electroweak scales. If one wants to extend the theory all the way to the Planck scale m_P (where gravity has to be included in the theory), then the natural value of the Higgs mass will be $m_P \sim 10^{19}$ GeV (!). Note that the divergence is not directly in the mass of the Higgs boson particle, but rather the quantum corrections affect the vev of the Higgs field, which in turn determines the masses of the electroweak bosons (and of course the Higgs mass itself). Fermion masses are protected by the chiral symmetry from diverging with Λ, but the introduction of an elementary scalar with a non-zero vev makes the theory extremely sensitive to the largest energy available. These corrections from loops with standard model particles alone (as in Fig. 9.1) destabilize the Higgs vev and push it to the ultraviolet cutoff of the theory. This is the Hierarchy problem. In other words, how come that $m_W \ll m_P$ (or equivalently, why $G_F \sim 1/v^2 \sim 1/m_W^2 \gg G_N = 1/m_P^2$)? This is

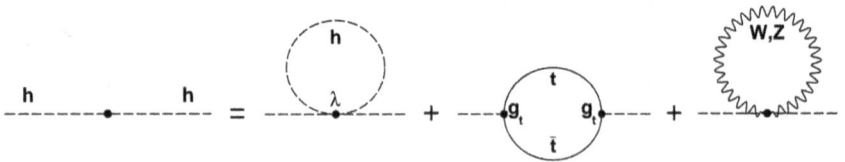

Figure 9.1: One-loop corrections to the Higgs boson mass from self-interaction (or interaction with any other possible scalars), fermion interactions (the dominant one is the top quark's), and gauge bosons interactions. The three diagrams are quadratically divergent and make the Higgs mass highly sensitive to the largest energy scale available.

a very deep problem, which comes back to the question of why gravity is so much weaker than the other forces.

One way to avoid this huge difference in energy scales is to postulate that the bare mass (the tree-level value of m_h^2) is close to the value of the quantum loop corrections and opposite in sign, such that they almost cancel yielding v^2. But this should strike you as rather unnatural: when subtracting two large numbers, their difference should naturally be of the same order, unless these numbers are almost equal to several significant digits. In this case, this would mean a cancellation of 36 orders of magnitude ($m_W^2 \approx -v^2 + m_P^2$)! This fine tuning (an extremely precise adjustment), between the bare mass and the quantum loop corrections, is more naturally explained if there is some (new) symmetry that allows to keep the bare mass close to the measured value and makes the quantum corrections small (within an order of magnitude of the the bare mass).

If we look more closely to the one-loop quantum corrections from Fig. 9.1, we can write out the different contributions from each diagram:

$$\Delta m_h^2 = \left[6\lambda - 6g_t^2 + \frac{1}{4}\left(9g^2 + 3g'^2\right)\right]\frac{\Lambda}{32\pi^2}. \qquad (9.2)$$

The first term arises from the Higgs self-interaction (λ) and is positive, the second term arises from the top-quark loop (g_t) and is negative. This is because of the spin-statistics theorem, which means that bosons have a positive contribution and fermions will have a negative contribution. To illustrate the problem of fine-tuning, if $\Lambda = 10$ TeV, the Higgs, top quark, and gauge bosons contributions to the mass term become: $(800 \text{ GeV})^2$, $-(1.5 \text{ TeV})^2$ and $(600 \text{ GeV})^2$, respectively.

It is clear from this picture that, in order to keep the bare mass close to the observed value and minimize the divergent terms from radiative corrections, any new physics must be bosonic in nature, to counteract the negative fermionic term with top quarks. The details are explained in Box 9.1.

Box 9.1 Radiative Corrections to the Mass Term

The heavy fermion loop (shown in the middle of Fig. 9.1) adds a radiative correction to the Higgs self-energy of the form:

$$
\begin{aligned}
\Delta m_h^2(f) &\approx -2N_f g_f^2 \int_0^\Lambda d^4k \left[\frac{1}{k^2 - m_f^2} + \frac{2m_f^2}{(k^2 - m_f^2)^2} \right] \\
&\propto -2N_f g_f^2 \left[\int_0^\Lambda \frac{k^3 dk}{k^2} + 2m_f^2 \int_0^\Lambda \frac{k^3 dk}{k^4} \right] \\
&\approx -2N_f g_f^2 \left[\int_0^\Lambda k \, dk + 2m_f^2 \int_0^\Lambda \frac{dk}{k} \right] \\
&\approx -2N_f g_f^2 \left[\Lambda^2 + 2m_f^2 \ln \Lambda \right],
\end{aligned}
\tag{9.3}
$$

where N_f is the number of fermions, g_f is the Yukawa coupling, k is the momentum of the particle (which changes as $d^4k = 2\pi k^3 dk$) and Λ is the cutoff energy, which is assumed to be much bigger than the masses.

The scalar loop (shown on the left of Fig. 9.1) contributes to the radiative correction as follows:

$$
\begin{aligned}
\Delta m_h^2(S) &\approx 2N_S g_S \int_0^\Lambda d^4k \left[\frac{1}{k^2 - m_S^2} + \cdots \right] \\
&\propto 2N_S g_S \left[\Lambda^2 + 2m_S^2 \ln \Lambda \right],
\end{aligned}
\tag{9.4}
$$

where S is a scalar particle with coupling g_S, and N_S is the number of scalar fields.

Any theory with new scalar particles will cancel the quadratic divergence if they have the same couplings, $g_f^2 = g_S$, and there is one scalar for each fermion, $N_S = N_f$. Furthermore, if all these particles had the exact same mass, $m_S = m_f$, then the two terms would cancel exactly. Supersymmetry (see Section 9.4) provides these extra scalars as the spin-0 partners of the top quark and the other fermions, and therefore elegantly solves the hierarchy problem.

9.3   Multiple Scalar Doublets

The standard model has the simplest possible Higgs sector with a single scalar field Φ, itself a complex $SU(2)$ doublet as introduced in Eq. (3.1). There are several reasons to suspect that Nature may have chosen a more diverse scalar sector, with more than one doublet. First, we have seen the very disparate masses of the bottom and top quarks at 5 and 172 GeV, respectively. In the standard model, all quarks receive their masses from the same doublet, while in a model with, for example, two doublets, it would be possible to have one set of Yukawa couplings for bottom-like quarks from one of the Higgs doublets, and another independent set of Yukawa couplings for the top-like quarks from the other Higgs doublet. This would be a more "natural" solution to their different values. Secondly, adding another scalar doublet adds several new scalar masses and introduces additional sources of CP violation, which in the standard model by itself seems to be not sufficient to explain the matter–antimatter asymmetry in the Universe. Adding a new scalar doublet naturally brings terms that explicitly or spontaneously break the CP symmetry. Finally, as will be discussed later in relation with supersymmetry, two Higgs doublets are necessary to give mass to different chiral multiplets.

Let's start by adding a second Higgs doublet to Eq. 3.1:

$$\Phi_1 = \sqrt{\frac{1}{2}} \begin{pmatrix} \phi_1 + i\phi_2 \\ \phi_3 + i\phi_4 \end{pmatrix}, \quad \Phi_2 = \sqrt{\frac{1}{2}} \begin{pmatrix} \phi_5 + i\phi_6 \\ \phi_7 + i\phi_8 \end{pmatrix}. \qquad (9.5)$$

These doublets have four degrees of freedom each (two positively charged components and two neutral components), and both acquire non-zero vevs, which spontaneously break the electroweak symmetry

$$(\Phi_1)_{\min} = \sqrt{\frac{1}{2}} \begin{pmatrix} 0 \\ v_1 \end{pmatrix}, \quad (\Phi_2)_{\min} = \sqrt{\frac{1}{2}} \begin{pmatrix} 0 \\ v_2 \end{pmatrix}, \qquad (9.6)$$

where v_1 and v_2 are real. If v_1 and v_2 were complex, any non-zero phase difference between the two vevs would spontaneously break

CP invariance. By making them real, we are assuming there is no CP violation in the scalar sector, just for simplicity. By putting v_1 and v_2 in the neutral (lower) components of Φ_1 and Φ_2, we avoid breaking the gauge symmetry of electromagnetism.

Out of the eight degrees of freedom from Eq. (9.5), three result in Goldstone bosons ($G^{\pm}$ and G^0), which provide the $W^{\pm}$ and Z boson with longitudinal polarizations, and hence mass, just like in the standard model. The remaining five degrees of freedom are realized as five physical massive scalars: two neutral CP-even h and H; a neutral CP-odd (pseudoscalar) A; and two charged $H^{\pm}$. We can write the multiplets in terms of these new scalars and the Goldstone bosons:

$$\Phi_1 = \sqrt{\frac{1}{2}} \begin{pmatrix} \sqrt{2}(H^+ \cos\beta + G^+ \sin\beta) \\ v\cos\beta - h\sin\alpha + H\cos\alpha + i(A\sin\beta - G^0\cos\beta) \end{pmatrix},$$

$$\Phi_2 = \sqrt{\frac{1}{2}} \begin{pmatrix} \sqrt{2}(H^+ \sin\beta - G^+ \cos\beta) \\ v\sin\beta + h\cos\alpha + H\sin\alpha + i(A\cos\beta + G^0\sin\beta) \end{pmatrix},$$

$$\tag{9.7}$$

where the angle α is introduced to diagonalize the mass matrix of the CP-even states h and H such that $m_H \geq m_h$. The angle β is defined as the ratio of the two doublets' vacuum expectation values: $\tan\beta = v_2/v_1$, with $\beta \in [0, \pi/2]$. Graphically, in terms of the potential, this means we can interchangeably use the coordinates (v_1, v_2), or the distance to the origin $v_{\mathrm{SM}} = \sqrt{v_1^2 + v_2^2}$ and the angle β. Here v_{SM} is the vev of the Higgs mechanism in the standard model (with only one complex doublet scalar), corresponding to $v_{\mathrm{SM}} \equiv v = 246$ GeV.

It is clear that this two-Higgs doublet model (2HDM) offers many new experimental challenges: new charged Higgs bosons, pseudoscalars and new couplings which lead to different decay modes and branching ratios. For example, the couplings of h and H to the gauge bosons now depend on factors of β and α, as shown in Table 9.1. In 2HDM, the sum of the squares of the h and H couplings to VV is equal to the square of the corresponding standard model Higgs coupling to VV.

Table 9.1: Vertex factors for h^0 and H^0 couplings to WW and ZZ. Each of these couplings is equal to the corresponding standard model hVV coupling in Table 3.1 except for the factors $\sin(\beta - \alpha)$ and $\cos(\beta - \alpha)$.

Interaction	Vertex factor $(-ig_{\mu\nu})$
hW^+W^-	$2\dfrac{m_W^2}{v_2}\sin(\beta - \alpha)$
hZZ	$2\dfrac{m_Z^2}{v_2}\sin(\beta - \alpha)$
HW^+W^-	$2\dfrac{m_W^2}{v_2}\cos(\beta - \alpha)$
HZZ	$2\dfrac{m_Z^2}{v}\cos(\beta - \alpha)$

To give mass to the fermions and gauge bosons, we can rewrite the kinetic term in Eq. (3.4) to include the new scalar doublet:

$$\mathcal{L}^{\text{kinetic}}_{\text{2HDM}} = (D_\mu^H \Phi_1)^\dagger (D_\mu^H \Phi_1) + (D_\mu^H \Phi_2)^\dagger (D_\mu^H \Phi_2)\,, \qquad (9.8)$$

Inserting the vevs of the Higgs doublets in this kinetic term, we find the mass terms for gauge bosons as follows:

$$
\begin{aligned}
m_W &= g\frac{\sqrt{v_1^2 + v_2^2}}{2} = g\frac{v_{\text{SM}}}{2}\,, \\
m_Z &= \sqrt{g^2 + g'^2}\frac{\sqrt{v_1^2 + v_2^2}}{2} = \sqrt{g^2 + g'^2}\frac{v_{\text{SM}}}{2}\,.
\end{aligned}
\qquad (9.9)
$$

For leptons, we can reuse the notation from Eqs. (3.16) and (3.17), to write out the kinetic term that gives them their mass now including the Yukawa couplings $g_1^{(e)}$ and $g_2^{(e)}$ from each doublet as follows:

$$-g_1^{(e)}\left[\bar{e}_L\Phi_1 e_R + \bar{e}_R\bar{\Phi}_1 e_L\right] - g_2^{(e)}\left[\bar{e}_L\Phi_2 e_R + \bar{e}_R\bar{\Phi}_2 e_L\right]\,. \qquad (9.10)$$

Reusing the notation of Eq. (3.20), we can now write the kinetic terms for down-type quarks as follows:

$$-g_1^{(d)}\left[\bar{d}_L\Phi_1 d_R + \bar{d}_R\bar{\Phi}_1 d_L\right] - g_2^{(d)}\left[\bar{d}_L\Phi_2 d_R + \bar{d}_R\bar{\Phi}_2 d_L\right], \qquad (9.11)$$

and for up-type quarks as follows:

$$-g_1^{(u)} \left[\bar{u}_L \Phi_1 u_R + \bar{u}_R \bar{\Phi}_1^c u_L \right] - g_2^{(u)} \left[\bar{u}_L \Phi_2 u_R + \bar{u}_R \bar{\Phi}_2^c u_L \right], \quad (9.12)$$

where we have used the scalar field Φ_i^c defined in Eq. (3.18).

This general 2HDM structure allows one to generate mass terms like the ones we obtained in Section 3.4 for the standard model. For example, for the down-type quark mass matrix, we now get the following:

$$\mathcal{M}^{(d)} = \left(g_1^{(d)} \frac{v_1}{\sqrt{2}} + g_2^{(d)} \frac{v_2}{\sqrt{2}} \right), \quad (9.13)$$

where the Yukawa couplings g_i are complex 3×3 matrices, so $\mathcal{M}^{(d)}$ can be diagonalized as in the standard model. In the standard model, when we diagonalize the mass matrices of the fermions, we are simultaneously diagonalizing the interaction terms between the Higgs boson and the quarks. When we try to do that with two different matrices g_1 and g_2 for each fermion type, it turns out that in general we cannot diagonalize the interaction terms automatically by diagonalizing the mass matrix. Having non-diagonal terms in the mass matrix creates terms, for example, of the form $A^0 \bar{s} d$, which would lead to a tree-level contribution to $K - \bar{K}^0$ mixing, as in Fig. 9.2. These processes are known as flavor-changing neutral currents (FCNCs) and are heavily constrained by experiments.

One could avoid these FCNC-inducing terms by setting many elements of the matrices g_1 and g_2 to zero (or making them

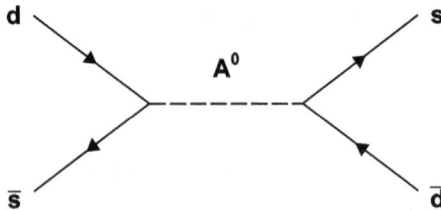

Figure 9.2: A sample Feynman diagram for $K^0 - \bar{K}^0$ mixing at tree-level, mediated by a pseudoscalar Higgs boson A^0 with flavor-changing couplings to down-type quarks.

small enough). This of course is not very elegant, so as usual when a model has terms we want to get rid of, we can require our theory to be invariant under a new discrete symmetry, such that, for example:

$$Z_2\Phi_1 = -\Phi_1, \quad Z_2\Phi_2 = +\Phi_2. \tag{9.14}$$

This Z_2 transformation gives Φ_1 a negative "charge" and leaves Φ_2 unchanged. We can also extend this symmetry to the right-handed fermions $Z_2 u_R = -u_R$, and only allow that one doublet couples to each type of fermion. This automatically removes all FCNCs. In this example, the terms in Eqs. (9.10)–(9.12) involving Φ_2 and $\bar{e}_R$ or $\bar{d}_R$, and the term involving Φ_1 and $\bar{u}_R$, are invariant under the Z_2 symmetry and hence allowed, while the other terms are forbidden. This method of avoiding FCNC by introducing a new discrete symmetry Z_2 is also called "natural flavor conservation".

If we ignore the neutrino masses, it can be shown that there are only four unique ways to assign the Z_2 charges for the right-handed fermions and the Higgs doublets. These different choices define four "types" of 2HDM and are displayed in Table 9.2.

For example, for Type II 2HDM, the right-handed up-type quarks couples to Φ_2, while the right-handed down-type quarks

Table 9.2: The four different types of 2HDMs with a symmetry that leads to natural flavor conservation, and the charges they assign to each field: the doublets Φ_1 and Φ_2, and the right-handed fermions u_R (up-type quarks), d_R (down-type quarks) and e_R (leptons). There are three generations of each field. By convention, the u_R always couples to Φ_2. A positive charge of a fermion it couples to Φ_2, and a negative one means it couples to Φ_1. Type X is often called *lepton-specific* and type Y is often called *flipped*.

Model	Φ_1	Φ_2	u_R	d_R	e_R
Type I	−	+	+	+	+
Type II	−	+	+	−	−
Type X	−	+	+	+	−
Type Y	−	+	+	−	+

and charged leptons couple to Φ_1. The Yukawa Lagrangian then becomes

$$\mathcal{L}_{\text{Yukawa}} = -g_1^\ell \bar{e}_L \Phi_1 e_R - g_1^{(d)} \bar{d}_L \Phi_1 d_R - g_2^{(u)} \bar{u}_L \Phi_2^c u_R + \text{h.c.} \quad (9.15)$$

The Yukawa couplings g_f are then given by the following:

$$g_u = \frac{\sqrt{2}m_u}{v_{\text{SM}}} \frac{1}{\sin\beta}, \quad g_{d,\ell} = \frac{\sqrt{2}m_{d,\ell}}{v_{\text{SM}}} \frac{1}{\cos\beta}. \quad (9.16)$$

These models, and other more exotic extensions including CP violation or alternative ways to reduce FCNCs, have been studied in the literature and are known to produce very diverse phenomenologies, which are under study by the experiments at the LHC. Not all the possible models may be probed at the LHC, but most of the parameter space should be accessible.

9.4 Supersymmetry and Its Breaking

The local gauge symmetries in the standard model, like SU(3) or U(1), are bosonic: their charges "q" (baryon number, electric charge, etc.) are scalar and transform particles with the same spin inside a multiplet. For example, the usual local gauge transformation $\alpha(x)$ in electrodynamics transforms an electron field: $\psi \to e^{iq\alpha(x)}\psi$ into another electron field with electric charge $q = -1$ and of course the same spin. The global symmetries of the Poincaré group are defined by the energy–momentum operator $P_\mu = i\partial_\mu$ and the Lorentz transformations $M_{\mu\nu} = i(x_\mu \partial_\nu - x_\nu \partial_\mu) + \frac{i}{4}[\gamma^\mu, \gamma^\nu]$. These are the generators of four space–time translations, three rotations and three boosts, and are also bosonic. Coleman and Mandula found out in 1967 that, indeed, when we try to combine internal symmetries like SU(3) or SU(2) with external Lorentz symmetries, in order to avoid scattering amplitudes equal to zero, all the bosonic generators must commute with the generators of the Poincaré group (P_μ and $M_{\mu\nu}$). But what about fermionic generators? Those can escape this restriction and produce a sensible quantum field theory that is non-trivial but possible. Let's introduce then a fermionic operator Q that transforms a fermion $|f\rangle$ into a boson $|b\rangle$, and vice versa:

$$Q|b\rangle = |f\rangle, \quad Q|b\rangle = |f\rangle. \tag{9.17}$$

These operators Q are fermionic and therefore must have a spinor index Q_α when acting on fields. The main role of Q_α is to change the spin of these fields so it will necessarily have to mix into the Poincare group of translations and rotations:

$$[Q_\alpha, \phi] \sim \psi, \quad \{Q_\alpha, \psi\} \sim \partial_\mu \phi + m\phi + g\phi^2 + \cdots \tag{9.18}$$

In fact, Haag, Lopuszanski and Sohnius in 1975 proved that the only possibility to extend the commutations of the Poincaré group is for these operators to obey the following algebra:

$$\{Q_\alpha, \bar{Q}_\beta\} = 2\sigma^\mu_{\alpha\beta} P_\mu, \tag{9.19}$$

where the labels α and β are spinor indices taking the values 1 and 2, the bar denotes conjugation. It is worth noting that non-relativistic analogs of this algebra are realized in Nature in nuclear, atomic and condensed matter physics. Thus, the appeal of searching for the last fundamental symmetry of Nature that is not forbidden but has not been observed yet!

This is supersymmetry. If we impose this new symmetry in the Lagrangian, then every fermion must have a bosonic partner, and vice versa: we are doubling the number of particles while keeping the same number of coupling constants. It is apparent now how supersymmetric theories solve the hierarchy problem: by having equal number of bosons and fermions, the residual loop corrections to the Higgs mass term in Eq. (9.2) are kept small because fermionic loops cancel out bosonic loops:

$$\Delta m_h^2 \sim \mathcal{O}\left(\frac{\alpha}{\pi}\right)(m_b^2 - m_f^2). \tag{9.20}$$

As long as the mass of the supersymmetric partner bosons b and fermions f are similar ($|m_b^2 - m_f^2| \lesssim 1$ TeV ^{2}), the radiative corrections should be below the observed Higgs mass. Supersymmetry provides the (new) bosons needed to oppose the dominant effect of the top quark (with its very large mass) in the loops.

It follows from the supersymmetric algebra that *supermultiplets* will consist of massless particles with spins differing by half a unit. Thus, fermions and bosons can be represented together inside the chiral, gauge and gravity supermultiplets:

$$\begin{pmatrix} f \text{ (fermion, } J = \tfrac{1}{2}) \\ \tilde{f} \text{ (sfermion, } J = 0) \end{pmatrix}, \quad \begin{pmatrix} b \text{ (gauge boson, } J = 1) \\ \tilde{b} \text{ (gaugino, } J = \tfrac{1}{2}) \end{pmatrix}, \quad \begin{pmatrix} G \text{ (graviton, } J = 2) \\ \tilde{G} \text{ (gravitino, } J = \tfrac{3}{2}) \end{pmatrix}.$$

We denote the superpartners with a tilde above the particle name. The superpartner naming convention is to add a prefix "s" to the fermion name and a suffix "ino" to the boson name, e.g., electron → selectron; W → Wino. Given this structure, could any of the existing fermions in the standard model be paired with the existing bosons to form these supermultiplets? No, for the simple reason that their internal quantum numbers do not match. For example, leptons have lepton number $L = 1$, but bosons have $L = 0$, thus they cannot be part of the same supermultiplet. This is why supersymmetry implies the existence of new particles (the superpartners) for all known particles: *squarks* and *sleptons* with spin 0, and the gauginos (Bino, Wino and gluino) and Higgsino with spin-1/2. As we will see shortly, instead of the photino and Zino, we typically only deal with Bino and Winos, which are expected to have masses larger than the electroweak scale. In supersymmetric models, there are only three generations of spin-1/2 quarks and leptons (no right-handed neutrino) as in the standard model. The left- and right-handed chiral fields belong to chiral super fields together with their spin-0 partners, the squarks and sleptons:

$$\hat{Q}, \hat{u}_R, \hat{d}_R, \hat{L}, \hat{\ell}_R.$$

The trick we used in Eq. (3.18) by using the complex conjugate of the Higgs field to give mass to the up-type quarks, does not survive in supersymmetric theories because a term of the form $\phi\phi^c$ is not allowed in the Lagrangian. In supersymmetry, two separate

Higgs doublets with opposite hypercharges are needed to give mass to the up- and down-type quarks. Hence, such a theory belongs to the Type II class of models in 2HDM, as described in Table 9.2. As it was the case for 2HDM, supersymmetry provides five different Higgs bosons after electroweak symmetry breaking. With two Higgs doublets, Φ_1 (sometimes called H_d) that gives mass to down-type quarks and Φ_2 (sometimes called H_u) that gives mass to up-type quarks, the supersymmetric Lagrangian can now include a term of the form $\mu H_u H_d$, where μ is called the mass-mixing parameter, analogous to the Higgs mass m_h in the standard model.

Just like in the standard model, once SU(2) × U(1) is broken, the weak eigenstates (Binos, Winos and Higgsinos) can mix to form mass eigenstates. The neutral higgsinos ($\tilde{H}_u^0$ and $\tilde{H}_d^0$) and the neutral gauginos ($\tilde{B}$ and $\tilde{W}^0$) combine to form four *neutralinos*: $\tilde{\chi}_{1,2,3,4}^0$, numbered according to their mass, from lightest to heaviest. The neutralinos are Majorana fermions: each particle is its own antiparticle. Similarly, the charged higgsinos ($\tilde{H}_u^+$ and $\tilde{H}_d^-$) and winos ($\tilde{W}^\pm$) mix to form four charged mass eigenstates called *charginos*: $\tilde{\chi}_{1,2}^\pm$.

Another consequence of supersymmetric models is the introduction of new terms in the superpotential that violate lepton- and baryon-number while still being consistent with gauge symmetries:

$$W_{\rm RPV} = \lambda_{ijk}\hat{L}_i\hat{L}_j\hat{\bar{\ell}}_{Rk} + \lambda'_{ijk}\hat{L}_i\hat{Q}_j\hat{\bar{d}}_{Rk} + \lambda''_{ijk}\hat{\bar{u}}_{Ri}\hat{\bar{d}}_{Rj}\hat{\bar{d}}_{Rk}, \qquad (9.21)$$

where i, j, k are generation indices of the fermions. The first term violates L, allowing for example this decay: $\bar{\ell} \to \ell + \ell$; the second term violates both B and L and could give rise to $\bar{q} \to q + \ell$; and the third term violates B and causes $\bar{q} \to \bar{q} + \bar{q}$. Any coupling that violates B and/or L is strongly constrained from nuclear physics, specially from the proton lifetime, direct searches, FCNCs and cosmology: if too large, they could wipe out the baryon asymmetry in the Universe. In the standard model, B–L conservation is never imposed by hand, it is rather an *accidental* outcome of the theory. In supersymmetry, one can impose the conservation of

B-L by creating a new multiplicative quantum number: R-parity, defined as

$$R = (-1)^{3(B-L)+2s}, \qquad (9.22)$$

for a particle of spin s. All standard model particles have R-parity of $+1$ while supersymmetric particles have R-parity of -1. Forcing the conservation of this new symmetry, several phenomenological consequences follow: supersymmetric particles (sparticles) are created in pairs ($pp \to \tilde{g}\tilde{g}$, $e^+e^- \to \tilde{\mu}^+ + \tilde{\mu}^-$), heavier sparticles decay to lighter ones ($\tilde{g} \to q\tilde{q}$, $\tilde{\mu} \to \mu\tilde{\chi}_1^0$), and the lightest sparticle (LSP) is stable because it has no allowed decay mode. This last point is another very attractive feature in these models since it provides neutral, weakly interacting, heavy and stable particles that could be candidates for dark matter: typically the lightest neutralino or the gravitino. If produced at colliders, the LSPs typically escape the detector without a detectable signal, thus behaving as heavy neutrinos. A common search strategy for such events is to use the momentum misbalance.

9.4.1 *Supersymmetry breaking, gravity and EWSB*

We have seen that supersymmetry predicts the existence of new particles with identical properties to their partners in the standard model except that their spins differ by $\frac{1}{2}$. So where are the selectrons and the photinos? Surely, we should have been able to find these particles if they had the same mass as electrons and photons. The fact that we have not implies that supersymmetry must be a broken symmetry. Unfortunately, there is no prediction for the mechanism of supersymmetry breaking. If explicit terms are introduced in the Lagrangian, they cannot induce quadratic divergences, or otherwise they would spoil the solution to the hierarchy problem. For example, dimensionless new Yukawa couplings or scalar quartic couplings would destroy the pattern of ultraviolet cancellations that keep the electroweak scale low. This leaves the only option of "soft" breaking terms with dimensions of mass. These are: scalar mass terms of the form $-m_{H_u}^2 H_u^\dagger H_u - m_{H_d}^2 H_d^\dagger H_d$; gaugino mass terms $-\frac{1}{2}[M_1\tilde{B}\tilde{B} + M_2\tilde{W}\tilde{W} + M_3\tilde{g}\tilde{g}]$; bilinear terms $-bH_uH_d$; and trilinear scalar interactions like $a_u\hat{u}_R\hat{Q}H_u$, where M_i, b and a_f are all parameters of the theory.

Another option is spontaneous supersymmetry breaking: letting a field acquire a vev. But there is no possible field in the Minimal Supersymmetric Standard Model (MSSM) that could be the primary source of supersymmetry breaking. Therefore, supersymmetry is usually thought to be spontaneously broken at some high energy scale M_S and therefore hidden, the same way that the full electroweak symmetry SU(2)×U(1) is hidden from very low-energy experiments. At energies $E > M_S$, the theory is supersymmetric and the sparticles are massless, while at energies $E < M_S$, the sparticles acquire a mass. The (new) hidden sector particles can be neutral with respect to the gauge groups of the standard model, and once supersymmetry is broken, it can be transmitted to the visible sector by some mechanism (perhaps with new messenger particles, or with standard model ones), as depicted in Fig. 9.3.

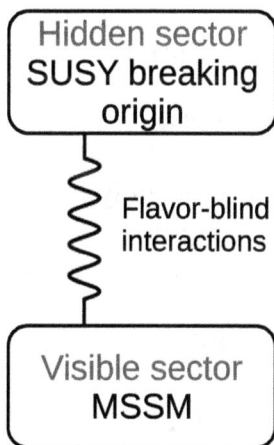

Figure 9.3: In the messenger paradigm, supersymmetry breaking occurs in a hidden sector (with large mass/energy scales) and is mediated down to the visible sector, for example, the MSSM, by interactions with either new particles (messengers) or standard model ones that link the two sectors.

The idea is that the hidden sector contains some field F that condensates acquiring a non-zero vev, F_0. Similarly to the Higgs mechanism, a Goldstone degree of freedom is needed, and this is provided by a fermion, the massless *Goldstino*, given that supersymmetry is a fermionic symmetry. If we assume that supersymmetry is a local symmetry (the spinor $\varepsilon(x)$ that allows to create superfields combining bosonic fields $\phi(x)$ and fermionic partners $\psi(x)$, such as $\phi(x) + \varepsilon(x)\psi(x)$, is space–time-dependent, and not just a constant), then gravity must be incorporated into the theory (see Box 9.2). The massless spin-3/2 gravitino only has two polarization states, the remaining two degrees of freedom required to become massive are acquired by eating the massless Goldstino spin-1/2 fermion. This superHiggs mechanism works just like the regular Higgs mechanism, where the gauge boson gains a longitudinal component (and therefore a mass) by eating the would-be Goldstone boson from spontaneously breaking a gauge symmetry. Here we achieve the same final result, a massive gravitino and a massless graviton, and local supersymmetry is thus successfully broken.

Box 9.2 Local Supersymmetry and Supergravity

Yet another incredible property of supersymmetry is that if we require it to be a local symmetry, it naturally involves gravity and opens the possibility of unifying all particle interactions and matter fields in Supergravity. Recall that in gauge theory one makes a space–time-dependent phase transformation $\epsilon(x)$:

$$\psi(x) \to e^{i\epsilon(x)}\psi(x). \tag{9.23}$$

A global symmetry would have a constant ϵ throughout space. When calculating the variation of the kinetic term: $\delta(i\bar{\psi}\gamma^\mu\partial_\mu\psi)$, this new field leads to a term of the following form:

$$-\bar{\psi}\gamma_\mu\psi\partial^\mu\epsilon(x), \tag{9.24}$$

(*Continued*)

(*Continued*)

which is then appropriately cancelled by the variation of the gauge interaction as follows:

$$\bar{\psi}(x)\gamma_\mu\psi(x)A^\mu(x) : \delta A^\mu(x) = \partial^\mu\epsilon(x). \tag{9.25}$$

Similarly, in supersymmetry we can make the supersymmetric "phase transformation" (in this case the spinor ε that flips each field) depend on space–time as follows:

$$\begin{aligned}
\delta\phi(x) &= \bar{\varepsilon}(x)\psi(x) \\
\delta\psi(x) &= -i\gamma_\mu\partial^\mu(\phi(x)\varepsilon(x)) + \cdots
\end{aligned} \tag{9.26}$$

The variation in the fermionic term then will include the following term:

$$\bar{\psi}\gamma_\mu\gamma_\nu\partial^\nu\phi\partial^\mu\varepsilon(x), \tag{9.27}$$

which has to be canceled by a new field Ψ^μ, a Majorana spinor field with spin $3/2$ that acts like a "gauge fermion", and has a coupling:

$$k\bar{\psi}\gamma_\mu\gamma_\nu\partial^\nu\phi\Psi_\mu(x) : \delta\Psi_\mu(x) = -\frac{1}{k}\partial_\mu\varepsilon(x). \tag{9.28}$$

This new field $\Psi_\mu(x)$ represents the gravitino. And voilà, by making supersymmetry local, we have brought gravity into the theory, and the gravitino plays the role of a gauge field in gauge theory!

This superHiggs mechanism is actually the only consistent way of breaking supersymmetry, just as the Higgs mechanism was the only consistent way of breaking gauge symmetry. The gravitino mass is then given by $m_{\tilde{G}} \sim F_0/M_P$, where F_0 is the energy scale at which supersymmetry breaking occurs in the hidden sector (like v^2 in the Higgs mechanism) and M_P is the Planck mass.

There are many possible interactions that can mediate supersymmetry breaking, for example: gravity (referred to as supergravity and sometimes abbreviated SUGRA), or the ordinary electroweak and QCD interactions (referred to as gauge-mediated supersymmetry breaking, or GMSB). Each case produces unique signatures and very different phenomenology: in SUGRA with R-parity conservation the neutralino is typically the LSP, while in GMSB, the LSP is the gravitino. The sparticle mass spectrum, and therefore what cascade decays become available at colliders, is highly sensitive to the many parameters (mass scales, mixing) of each model of supersymmetry breaking.

Finally, let's discuss how supersymmetry breaking includes and further explains electroweak symmetry breaking. In the Higgs mechanism, there is no explanation for why the Higgs mass term is negative, $m_h^2 < 0$, such that one obtains the Mexican hat potential needed for electroweak symmetry breaking, as show in Fig. 3.2. In supersymmetry, with so many new scalars, which one should be the one that acquires a non-zero vev? We know that $\langle H_u \rangle \neq 0$ breaks SU(2) × U(1), but it would be just as easy to have the stop quark acquire a vev, $\langle \tilde{t}_R \rangle \neq 0$, which would break color SU(3) but preserve SU(2). Remarkably, supersymmetry provides a natural explanation for why it has to be the up-type quark Higgs mass term that becomes negative, and not others. It has to do with the large value of the top quark Yukawa coupling. When radiative corrections are incorporated for the different mass terms in their renormalization group evolution, it turns out that the H_u term is positive at large energies, but becomes negative because the terms involving top-quark loops dominate and become negative as $m_{H_u}^2$ runs down to the electroweak energy scale. This is shown in Fig. 9.4. It is quite beautiful, supersymmetry provides a dynamical theory of SU(2)×U(1) spontaneous symmetry breaking simply based on the large top quark Yukawa coupling!

Quite depressingly for such a complete and elegant theory, however, direct searches for supersymmetry performed at LEP, the Tevatron and the LHC have all resulted in negative results. Some

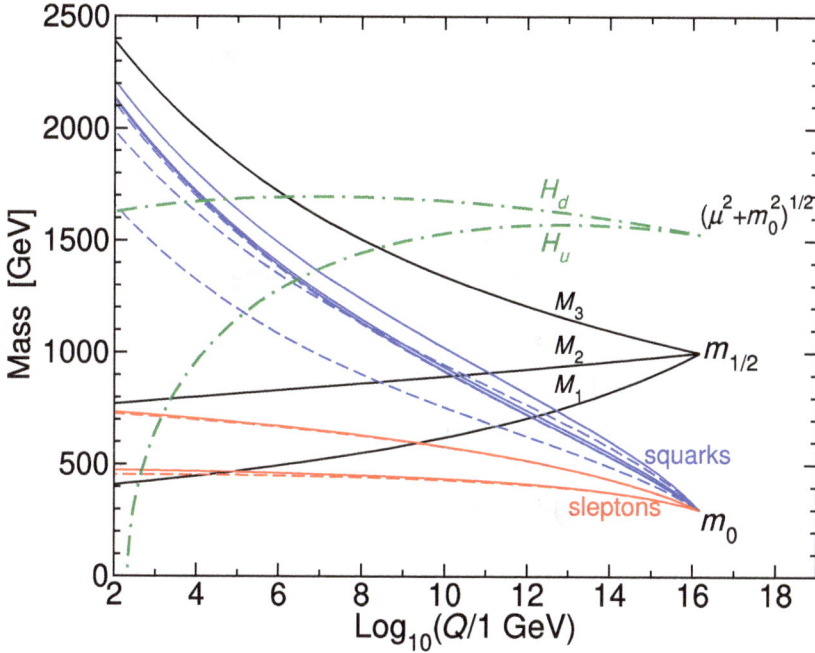

Figure 9.4: Renormalization group evolution of scalar and gaugino mass parameters in the MSSM with minimal-SUGRA boundary conditions imposed at $Q_0 = 1.5 \times 10^{16}$ GeV. The parameter $\mu^2 + m_{H_u}^2$ runs negative, provoking electroweak symmetry breaking. $M_{3,2,1}$ are the three complex gaugino Majorana mass parameters, associated with the SU(3), SU(2) and U(1) subgroups of the standard model (the gluino, wino and Bino masses), respectively. These are assumed to unify at the GUT scale into the gaugino common mass $m_{1/2}$. Similarly, the squarks and sleptons unify at a common scalar mass m_0 at the GUT scale.

simple, more natural models in SUGRA (and others) have been explored, and strict limits have been obtained for the mass of the sparticles in that framework: $m_{\tilde\chi_1^0} \gtrsim 300$ GeV, $m_{\tilde g} \gtrsim 2$ TeV and $m_{\tilde t_1} \gtrsim 1$ TeV. This has led to the following constraints in order to keep the electroweak scale at ~ 100 GeV within supersymmetry:

- $m_{H_u}^2 \sim -(100\text{–}300)$ GeV^2 at the weak scale;

- $|\mu| \sim 100\text{–}300$ GeV; and

- the stop quark radiative corrections to the μ parameter cannot be too large. The somewhat large value of $m_h = 125$ GeV requires either a large stop quark mass, or a large mixing of the weak eigenstates leading to a light mass eigenstate (which may be more natural).

It is important to note that the experimental limits quoted above can be evaded if we allow for a richer parameter space and relax some assumptions in the simplest models. Even the simpler models still have hundreds of parameters (sparticle masses, mixing angles, CP-violating phases, trilinear couplings, R-parity violating couplings, etc.) that emerge from the enlarged particle zoo and by parametrizing how supersymmetry may be broken. This theory is clearly elegant and appealing because it treats matter and interactions (fermions and bosons) in a unified way, but the cost is a very rich phenomenology and no shortage of new unknown parameters!

Suggested Reading for Chapter 9

[1] Sidney R. Coleman and J. Mandula. "All possible symmetries of the s matrix". *Phys. Rev.* 159 (1967) pp. 1251–1256.

[2] Rudolf Haag, Jan T. Lopuszanski, and Martin Sohnius. "All Possible Generators of Supersymmetries of the s Matrix". *Nucl. Phys. B* 88 (1975), p. 257.

[3] Aker, M. et al. "Improved upper limit on the neutrino mass from a direct kinematic method by Katrin". *Phys. Rev. Lett.* 123(22) 2019, p. 221802. arXiv: 1909.06048 [hep-ex].

[4] Fernando Quevedo, Sven Krippendorf, and Oliver Schlotterer. "Cambridge Lectures on Supersymmetry and Extra Dimensions" (Nov. 2010). arXiv: 1011.1491 [hep-th].

[5] G. C. Branco et al. "Theory and phenomenology of two-Higgs-doublet models". *Phys. Rept.* 516 (2012), pp. 1–102. arXiv: 1106.0034 [hep-ph].

[6] Stephen P. Martin. "A Supersymmetry primer". *Adv. Ser. Direct. High Energy Phys.* 18 (1998), pp. 1–98. arXiv: hep-ph/9709356.

[7] Hitoshi Murayama. "Supersymmetry phenomenology". *ICTP Summer School in Particle Physics.* Feb. 2000. arXiv: hep-ph/0002232.

[8] John R. Ellis. "Supersymmetry for Alp hikers". *2001 European School of High-Energy Physics.* Mar. 2002. arXiv: hep-ph/0203114.

[9] D. G. Cerdeno and C. Munoz. "An introduction to supergravity". *PoS CORFU98* (1998), p. 011.

Chapter 10

Putting Things in Perspective

10.1 Cosmological Role of Scalar Fields

To understand the cosmological role of scalar fields, we need to consider a simple primer from General Relativity (GR) — a homogeneous universe. The behavior of such a system is determined by the equation of state — the relationship between pressure, P, and density, ρ. Based on dimensional analysis it is easy to see that pressure is equal to density times c^2 times some dimensionless coefficient w:

$$P = w\rho c^2. \tag{10.1}$$

The value of w defines the equation of state. Cold matter, e.g., cosmic dust, does not exert any pressure, thus $w = 0$. For relativistic species, e.g., radiation, $w = 1/3$.[1] Of particular interest is the behavior of a substance with $w = -1$, which implies negative pressure.

To describe the expanding universe, let us define the comoving coordinates. Consider a stretchable piece of tartan cloth. The comoving coordinates, x are set by the grid of tartan, e.g., two buttons are sewn three squares from each other. If the cloth is stretched, this relationship does not change, even though the physical distance, r, measured in meters does. In three dimensions, this piece of cloth

---

[1] It is easy to derive considering photons bouncing off the wall at an angle θ. A factor of $1/3$ arises from averaging over the $\cos^2 \theta$.

turns into an expanding universe. The time-dependent scale factor $a(t)$ relates the comoving and physical coordinates as follows:

$$r(t) = a(t)x. \tag{10.2}$$

Typically, t is counted from the birth of the universe and it is assumed that the two coordinates are identical in present day universe $t = t_0$, hence $a(t_0) = 1$. In the case of a homogeneous universe, the comoving coordinates x of some object, e.g., a galaxy, do not change with the expansion, only the scale a does. Thus, differentiating this equation over time, we get the following:

$$\dot{r} = \dot{a}x = \frac{\dot{a}}{a}r, \tag{10.3}$$

which states that the velocity of expansion is proportional to the distance to the object, which is Hubble's law. In the local universe the relative expansion rate defines the Hubble's constant,

$$H_0 = \frac{\dot{a}}{a}(t_0). \tag{10.4}$$

The first law of thermodynamics, which is just the energy conservation, states that

$$dE = -PdV, \tag{10.5}$$

where the internal energy E can be related to density using Einstein's famous equation $E = mc^2 = \rho c^2 V$. Volume V can be expressed in the comoving coordinates as $V = a^3 x^3$. Hence, plugging in the equation of state, we get

$$d(\rho a^3) = -w\rho d(a^3), \tag{10.6}$$

where we canceled x^3 since it remains constant with the expansion. The solution to this differential equation has the form

$$\rho = \rho_0 a^{-3(w+1)}, \tag{10.7}$$

where we assume the boundary condition that at present time $\rho(a = 1) = \rho_0$. For cold matter with $w = 0$, this results in a very natural dependence of density on the scale $\rho = \rho_0 a^{-3}$, or inversely proportional to volume, which means that the enclosed mass remains constant. For radiation with $w = 1/3$, we get $\rho = \rho_0 a^{-4}$. This relationship is also easy to understand. With the expansion, not only does the photon number density go down as a^{-3} just like for cold matter but the wavelength also "stretches" linearly with a, so that the observed wavelength, λ_{obs}, is longer (redder) than the emitted wavelength λ as follows:

$$\lambda_{obs} = \frac{\lambda}{a(t)} = \lambda(z + 1), \tag{10.8}$$

where in the second equation we introduced the definition of the cosmological redshift, z. It is similar in its result to the classical Doppler effect, but it arises from the expansion of space itself, not the relative velocity between emitter and observer. A simple relation ties together the redshift, which is a physical observable, and the scale $a = (z + 1)^{-1}$. Therefore, as we set out to demonstrate, the energy of a single photon decreases inversely proportionally to a. And as a result, the energy density of radiation decreases as a^{-4}. For this reason, radiation only plays a major role in the expansion of the universe for very small value of a. Interestingly, since neutrinos are very light, they are treated as a relativistic species and as such they contribute on par with radiation to the evolution of the universe at these very early stages. Hence, the number of neutrino families (determined by LEP and SLC to be three, see Box 2.3) is important for the early universe expansion.

If w is equal to -1, Eq. (10.7) tells us that the density of such a substance remains constant with the expansion. Every new piece of volume that is "added" to the expanding universe has the same energetic value. This particular type of energy is related to the "cost of land", or the vacuum energy. We will refer to the vacuum energy density as ρ_V. In the context of a scalar field, it is related to the non-zero value of the potential at its minimum, as we have seen in the case of spontaneously broken symmetry. In the context of cosmology, vacuum energy is frequently referred to as the dark energy.

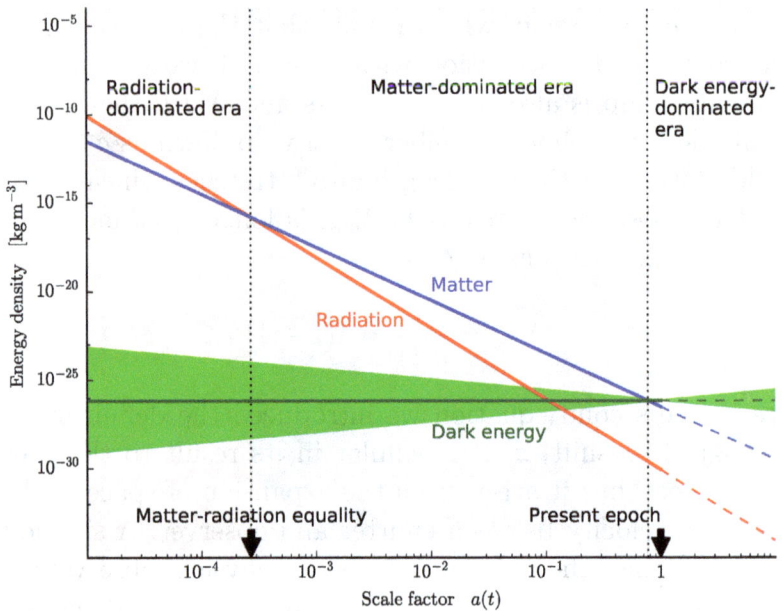

Figure 10.1: The evolution of the energy density of the main components of the universe. For dark energy, the green band represents an equation of state parameter $w = -1 \pm 0.2$, showing how a small change in the value of this parameter can give very different evolution histories for dark energy. The matter and radiation densities will keep decreasing as the universe expands (dashed lines).

The evolution of radiation, matter and dark energy density with the scale is demonstrated in Fig. 10.1. Radiation dominated in the early stages, until the universe expanded enough for matter density to become dominant at a redshift of $z \approx 3,600$, or when the universe was 47,000 years old. Because the matter density dropped as the scale factor increased, dark energy, whose density remained constant, began to dominate in the relatively recent past, at $z \approx 0.5$, when the universe was 9.8 billion years old. Incidentally, the Earth was already formed at that time. Now the universe is 13.8 billion years old, and the dark energy constitutes about 70% of the matter-energy budget.

For a homogeneous universe, the equations of GR are reduced to the following Friedmann equation:

$$\left(\frac{\dot{a}}{a}\right)^2 = \frac{8\pi G_N}{3}\rho - \frac{Kc^2}{a^2}, \tag{10.9}$$

where G_N is Newton's constant and K is the intrinsic curvature of space. $K > 0$ corresponds to a closed spherical universe, $K < 0$ to open hyperbolic space, and $K = 0$ is the critical case of a flat universe. For simplicity, we shall consider the case of a flat universe, especially since the observational data points to it. Combining Eqs. (10.4) and (10.9), and setting $a = 1$, we get that the value of Hubble's constant is determined by the present day energy density as follows:

$$H_0^2 = \frac{8\pi G_N}{3}\rho_0. \tag{10.10}$$

Let us consider the case when energy density is dominated by vacuum $\rho_0 = \rho_V = \frac{\Lambda}{8\pi G_N}$, where Λ is the cosmological constant. In this case, the Hubble constant is determined by the value of Λ as follows:

$$H_0^2 = \frac{8\pi G_N}{3}\rho_V = \frac{\Lambda}{3}. \tag{10.11}$$

If the energy density is constant, the solution to the differential equation Eq. (10.9) is

$$a(t) = \exp[H_0(t - t_0)], \tag{10.12}$$

which corresponds to the exponential expansion of the universe.

Plugging in the corresponding dependence of energy density on scale into Eq. (10.9) we get that the radiation-dominated universe expands proportionally to $\sim t^{1/2}$, while the matter-dominated universe follows the $\sim t^{2/3}$ expansion (see Table 10.1).

We see how a cosmological constant due to non-zero vacuum energy provides the currently observed accelerated cosmological

Table 10.1: The evolution of the main cosmological components. Radiation includes photons and relativistic neutrinos. The matter (or sometimes, dust) component includes cold dark matter, baryons and non-relativistic neutrinos. Vacuum energy and inflation depend on the Hubble parameter H.

Component	w	$\rho = a^{3(1+w)}$	$a(t) = t^{2/3(1+w)}$
Radiation	$1/3$	$\sim a^{-4}$	$\sim t^{1/2}$
Matter	0	$\sim a^{-3}$	$\sim t^{2/3}$
Vacuum energy	-1	constant	$\sim e^{Ht}$
Inflation	$\to -1$	$\frac{1}{2}\dot{\phi}^2 + V(\phi)$	$\sim e^{Ht}$

expansion, as established by the observation of Type Ia supernovae and the cosmic microwave background (CMB). This constant, however, does not explain a super-accelerated expansion, referred to as *inflation*, in very early times, as is needed to explain why the universe is flat, and why causally disconnected regions in the observable universe appear to have the same temperature (the so-called *horizon problem* of Big Bang cosmology). The current accelerated expansion seems to have started recently, in cosmological terms, while any period of accelerated expansion in the very early universe must be made to come to an end before all the particles appear in the primordial soup. Therefore, the vacuum energy driving these different periods of expansion must be dynamic. It should come as no surprise at this point that a scalar field with a non-zero vev can be used to provide a time-dependent "cosmological constant". This field should have the quantum numbers of the vacuum, and have a non-zero ground state potential that explains the vacuum energy density: something quite similar to what the Higgs field provided to break the electroweak symmetry. So, once again, scalar fields can play a crucial role in the fate of the universe.

Let's consider a scalar field ϕ minimally coupled to gravity. We don't know what potential $V(\phi)$ this field may have, but we know that in quantum field theory, the addition of a constant to the

Lagrangian density does not change the Euler–Lagrange equations and thus does not affect the physical observables. In other words, the potential given by Eq. (3.2) can be redefined so that its minimum has an arbitrary value as illustrated in Fig. 10.2. At the same time in GR, gravity is sensitive to the total value of the potential since it couples to the energy–momentum tensor. The universe starts in a state of unbroken symmetry with $\phi_{\min} = 0$ and $V(\phi_{\min}) = V_0$. Then it cools down and develops a minimum displaced from zero: $\phi_{\min} = \phi_1$ and $V(\phi_1) = V_1$. The universe then "rolls off" from the state with very high value of the potential energy (and thus ρ_V) to a state with a small, but non-zero value.

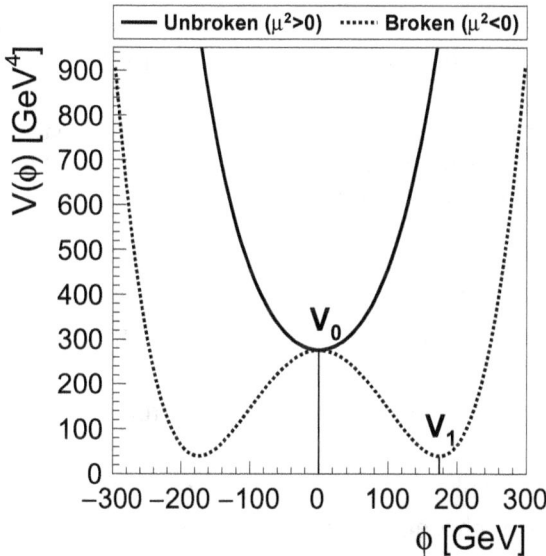

Figure 10.2: Electroweak potential before (solid line) and after (dashed line) electroweak symmetry breaking. The potential energy associated with the ground state before the phase transition is V_0, and after it is V_1. Quantum field theory is insensitive to V_1, only to $V_0 - V_1$, while GR "feels" the total value of the potential energy. Since $V_0 \gg V_1$, the rate of the expansion is much larger before the EW phase transition. The residual value of V_1 determines the present day value of the cosmological constant.

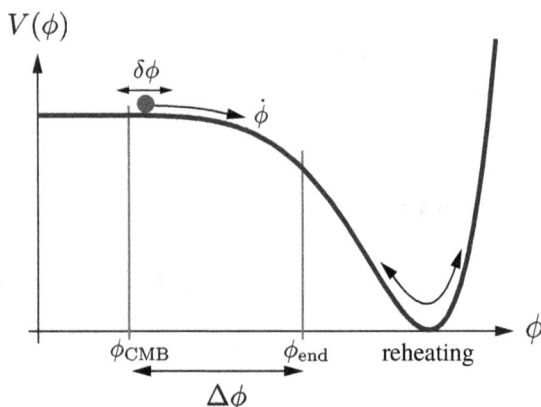

Figure 10.3: Example of an inflaton potential and the slow-roll mechanism. Acceleration occurs when the potential energy of the field $V(\phi)$ dominates over its kinetic energy $\frac{1}{2}\dot{\phi}^2$. Inflation ends at ϕ_{end} when the kinetic energy has grown to become comparable to the potential energy $\frac{1}{2}\dot{\phi}^2 \equiv V$. CMB fluctuations are created by quantum fluctuations $\delta\phi$ about 60 e-folds before the end of inflation. At reheating, the energy density of the inflaton is converted into radiation and the observed standard model particles.

This explains a fast exponential expansion at the very early stages of the universe history and a slow, but still exponential expansion in the present day. In between, the universe went through the radiation-dominated era followed by the matter-dominated epoch (see Fig. 10.1).

The exact form of the potential does not have to follow the parametrization chosen in Eq. (3.2), which can be viewed as a Taylor expansion around the minimum. As long as the function can develop a minimum displaced from zero, it will generate the spontaneous symmetry breaking. Whether the scalar field that causes inflation is the Higgs field, or yet another member of the scalar community, called the *inflaton*, is subject to a debate in the theoretical community (see Fig. 10.3 for an example of slow-roll inflation driven by inflaton scalar field). One thing is sure, the exact role of scalar fields might reach far beyond the generation of masses to gauge bosons and fundamental fermions.

10.2 On the Philosophical Meaning of Gauge Invariance

The principle of gauge invariance was noticed fairly early in the history of physics. It has to do with the freedom in the definition of scalar and vector potentials in the theory of electromagnetism. Similarly, in quantum mechanics the wave function is defined up to a common phase factor. That is, it can be multiplied by $e^{i\alpha}$, where α is an arbitrary constant, without changing the experimentally observable results. With the development of quantum field theory, it became clear that the definitions are related. Requiring that the Lagrangian remains invariant, even if the phase $\alpha(x)$ is a function of the position in space, leads to the generation of the gauge fields. This is true for electromagnetic, strong and weak fields. In other words, gauge fields communicate the information about the changed calibration (gauge) in different points in space. As we have seen, the demand of local gauge invariance in a non-Abelian group results in self-interaction of the gauge fields. This prediction was experimentally observed as triple and quadruple gauge boson vertices. Massive gauge bosons would have broken the gauge invariance of the Lagrangian. To prevent this from happening, a scalar field is introduced with an associated potential with a non-zero vacuum expectation value. As a result, the entire world of the elementary particles acquired mass. It is puzzling why Nature decided to go to such great lengths to conceal the phase of the wave function.

Suggested Reading for Chapter 10

[1] Ivan Debono and George F. Smoot. "General Relativity and Cosmology: Unsolved Questions and Future Directions". *Universe* 2.4 (2016). ISSN: 2218-1997. URL: https://www.mdpi.com/2218-1997/2/4/23.

[2] Peter W. Graham, David E. Kaplan, and Surjeet Rajendran. "Cosmological Relaxation of the Electroweak Scale". *Phys. Rev. Lett.* 115.22 (2015), p. 221801. arXiv: 1504.07551 [hep-ph].

[3] Gia Dvali. "Cosmological Relaxation of Higgs Mass Before and After LHC and Naturalness" (Aug. 2019). arXiv: 1908.05984 [hep-ph].

[4] Daniel Baumann. "Inflation". *Theoretical Advanced Study Institute in Elementary Particle Physics: Physics of the Large and the Small*. 2011, pp. 523–686. arXiv: 0907.5424 [hep-th].

[5] Javier Rubio. "Higgs Inflation". *Frontiers in Astronomy and Space Sciences* 5 (2019). ISSN: 2296-987X. URL: https://www.frontiersin.org/article/10.3389/fspas.2018.00050.

Relevant Reviews

A.1 Kinematics

A.1.1 *Covariant notation*

Lorentz vectors are labeled a_μ, where $\mu = 0, 1, 2, 3$ is the time–space index. The covariant product of two vectors a and b is defined as

$$ab = g_{\mu\nu}a^\mu b^\nu = a^0 b^0 - a^1 b^1 - a^2 b^2 - a^3 b^3, \qquad (\text{A.1})$$

where

$$g_{\mu\nu} = \begin{pmatrix} 1 & 0 & 0 & 0 \\ 0 & -1 & 0 & 0 \\ 0 & 0 & -1 & 0 \\ 0 & 0 & 0 & -1 \end{pmatrix}. \qquad (\text{A.2})$$

is the Minkowski metric. We assume summation over repeating indices. We use Greek indices for Lorentz vectors, labeled as a, and Latin indices for three-dimensional vectors, labeled as $\vec{a}$.

An example of a covariant product is the $p^2 = E^2 - \vec{p}^2 = m^2$ for a particle with mass m, energy E, and momentum $\vec{p}$. It can also be defined for a system of particles, with a Lorentz momentum p. In this case m is referred to as the invariant mass of the system.

In collider experiments, a coordinate system is typically defined such that the z axis is directed along the beam direction, the x axis points to the center of the collider ring, and the y axis is vertically upwards. The polar angle θ is the angle of particle's momentum

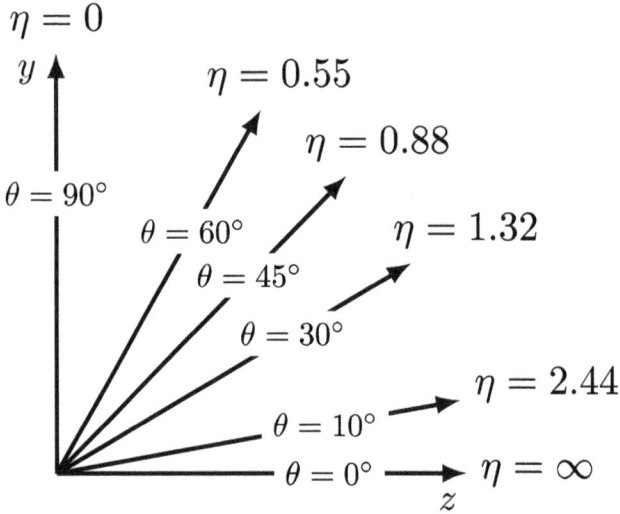

Figure A.1: The correspondence between pseudorapidity η and polar angle θ.

with the z axis, and azimuthal angle ϕ is the angle in the transverse to the beam plane. In hadron colliders, a useful variable is pseudorapidity $\eta = -\ln[\tan(\theta/2)]$, as shown in Fig. A.1. In the ultrarelativistic limit $(v \to c)$, pseudorapidity approximates the true rapidity defined as

$$y = \frac{1}{2} \ln \frac{E + p_z}{E - p_z} = \tanh^{-1} \frac{p_z}{E}, \tag{A.3}$$

where p_z is the component of the momentum parallel to the beam axis. Rapidity is useful because it transforms linearly under a boost in the z direction, and hence differences in rapidity are invariant, and so is the shape of the rapidity distribution dN/dy for particle production. Pseudorapidity is useful because it can be calculated without knowing the momentum and total energy.

A.1.2 *Observed events, backgrounds, significance*

The number of events observed in an experiment is

$$N_{\text{obs}} = \sigma \mathcal{L} \varepsilon, \tag{A.4}$$

where σ is the cross section of the physics process, measured in barn $= 10^{-24}$ cm^2, $\mathscr{L}$ is the integrated luminosity delivered by the collider, measured in barn^{-1}, and ε is the efficiency of the experiment to detect these events. Hence, the experimentally measured (differential) cross section, which could be a function of a certain variable x, e.g., momentum of the particle, is evaluated as

$$\frac{d\sigma(x)}{dx} = \frac{dN_{\text{obs}}(x)}{dx}\frac{1}{\mathscr{L}\varepsilon}. \tag{A.5}$$

The experimentally measured cross sections can be compared to the theoretical prediction to verify the validity of the theory.

If a certain type of events can be produced via different physics processes, the observed number is a sum of the events expected from each source. Typically, the number of observed events is the sum of events due to the process of interest (the signal, N_s) and background events (N_b). Events observed in an experiment are

$$N_{\text{obs}} = N_s + N_b. \tag{A.6}$$

Care should be taken if the initial and final states for signal and background are identical. Then, the joint cross section must be evaluated, by summing up the amplitudes for the corresponding processes.

The number of both signal and background events scale linearly with the integrated luminosity. The significance of the signal is defined as the probability of making an observation that is at least as inconsistent with the null hypothesis as the observation actually made. When optimizing an analysis, it is customary to use an approximated expression of the significance (sometimes called the sensitivity) that does not involve complicated statistical machinery, and relies only on counting events. Typically:

$$S_s = \frac{N_s}{\Delta N_b} \approx \frac{N_s}{\sqrt{N_b}}, \tag{A.7}$$

where ΔN_b is the uncertainty on the background. For counting experiments with Poisson distribution, the standard deviation of the number of background events[1] is given by $\sqrt{N_b}$, which is proportional to $\sqrt{\mathcal{L}}$. For this reason, the signal significance scales as $S_s \propto \mathcal{L}/\sqrt{\mathcal{L}} = \sqrt{\mathcal{L}}$.

This definition of significance, however, does not behave well when N_b is very small, or if N_s is not much smaller than N_b. In that case, and ignoring the uncertainty on the background ΔN_b, it is recommended to use the following:

$$S_s = \sqrt{2\left[(N_s + N_b) \times \ln\left(1 + \frac{N_s}{N_b}\right) - N_s\right]}. \qquad (A.8)$$

For small signals it is necessary to accumulate significant integrated luminosity. An agreed upon convention in particle physics is that evidence for a certain process can be claimed when the significance is equal to 3 (referred to as 3σ-level). Discovery corresponds to 5σ-level. This nomenclature is used even if the fluctuations are not described by Gaussian statistics. In this case the probability for background events to fluctuate so that they mimic the signal (called p-value) must be at the same level, as it would be for 5 (3) standard deviations in case of Gaussian statistics. Of course, if you want to claim that you have observed a process where energy is not conserved, or you have found a particle that moves faster than the speed of light, it may be prudent to require a much higher significance than 5 standard deviations! The 3 and 5σ thresholds are just conventions in the field based on past experience.

A.1.3 *Cross section*

According to Fermi's Golden rule, the differential cross section of two particles A and B with four-momenta p_i ($i = A, B$), going to the final state of N particles with four-momenta p_f ($f = 1, ..., N$),

---

[1]Typically, for $N \geq 20$ the Poisson distribution can be approximated by a Gaussian distribution.

is given by

$$d\sigma = \frac{|\mathcal{M}|^2}{F} dQ, \tag{A.9}$$

where $\mathcal{M}$ is the matrix element of the process $A + B \to N$,

$$F = 4 \left[(p_A p_B)^2 - (m_A m_B)^2 \right]^{1/2}, \tag{A.10}$$

is the initial flux and dQ is the Lorentz invariant phase space (LIPS) of the final state:

$$dQ = (2\pi)^4 \delta^4 \left(\sum_i p_\mu^i - \sum_f p_\mu^f \right) \prod_{f=1}^{N} \frac{d^3 p_f}{(2\pi)^3 2 E_f}. \tag{A.11}$$

If masses of the initial state particles can be neglected, $F = 2s$ in the center-of-mass. For 2-to-2 process, e.g., $A + B \to C + D$, in center-of-mass the momenta of the final state particles have the same magnitude p_f and are opposite in direction. Then the differential cross section can be expressed as

$$\frac{d\sigma}{d\Omega} = \frac{|\mathcal{M}|^2}{64\pi^2 s} \frac{p_f}{p_i}, \tag{A.12}$$

where $d\Omega = d(\cos\theta) d\phi$ is the solid angle of the final product C. If in addition the masses of the final state particles are the same $m_C = m_D = m_f$, their energies are then equal $E_f = \sqrt{s}/2$. We can introduce the relativistic factor $\beta^2 = 1 - 4m_f^2/s$ for final state particles. Then

$$\frac{d\sigma}{d\Omega} = \frac{|\mathcal{M}|^2}{64\pi^2 s} \beta. \tag{A.13}$$

Here we explicitly used that the absolute value of the initial state momentum p_i is equal to the energy of the final state particles E_f. In the ultra relativistic limit (neglecting final state masses), the relativistic factor $\beta \to 1$. It is worth noting that matrix element $\mathcal{M}$ is proportional to β^l, where l is the orbital quantum number of the final state. So, if the reaction goes in s-wave mode, the cross section depends on β^1; and if in p-wave, on β^3. These dependencies are particularly important near the production threshold.

A.1.4   *Decay widths*

Partial differential decay width of particle A with energy E_A is given by

$$d\Gamma = \frac{|\mathcal{M}|^2}{2E_A} dQ, \qquad (A.14)$$

where the factor of $2E_A$ comes from the normalization of the initial state wave function and dQ, the Lorentz invariant phase space of the final state, has the same expression as in Eq. (A.11). In its center-of-mass, the energy of particle A is equal to its mass $E_A = m_A = \sqrt{s}$. If particle A decays to two particles C and D in center-of-mass, their momenta have the same magnitude p_f and are opposite in direction. Performing the integration we get

$$\frac{d\Gamma}{d\Omega} = \frac{p_f |\mathcal{M}|^2}{32\pi^2 m_A^2}. \qquad (A.15)$$

If in addition the masses of the final state particles are the same $m_C = m_D = m_f$, then

$$\frac{d\Gamma}{d\Omega} = \frac{\beta |\mathcal{M}|^2}{64\pi^2 m_A}. \qquad (A.16)$$

If the decay is isotropic, the integration over solid angle yields 4π and the partial width is equal to

$$\Gamma = \frac{\beta |\mathcal{M}|^2}{16\pi m_A}. \qquad (A.17)$$

Similarly to the cross section, the decay width depends on the relativistic factor as β^{2l+1}, where l is the orbital quantum number.

A.1.5   *Breit–Wigner formula*

The cross section for production of a short-lived particle (resonance) of mass m and total width Γ in collisions of particle–antiparticle of type i at the center-of-mass energy $\sqrt{s}$

with subsequent decay to particles of type f is described by the Breit–Wigner formula:

$$\sigma = \frac{4\pi(2J+1)}{s} \frac{\Gamma_i \Gamma_f}{(\sqrt{s}-m)^2 + \Gamma^2/4}, \tag{A.18}$$

where Γ_i (Γ_f) is the partial width of the particle decay to the initial (final) state particles, and J is the spin of the resonance.

A.2 Spin 1/2 Particles

Let us consider the case for $s = 1/2$ fermion, such as a lepton, a neutrino, or a quark.

A.2.1 *Properties of the SU(2) symmetry group*

In mathematics, the special unitary group of degree n, denoted SU(n), is the Lie group of $n \times n$ unitary matrices with determinant 1. The lowest non-trivial value of n is 2, forming SU(2). In physics, this is a rotation group with the generators being the operators of spin $\hat{s} = \sigma_i/2$, $i = 1, 2, 3$, where σ_i are 2×2 Pauli matrices. They are traceless and unitary:

$$\sigma_1 = \begin{pmatrix} 0 & 1 \\ 1 & 0 \end{pmatrix}, \quad \sigma_2 = \begin{pmatrix} 0 & -i \\ i & 0 \end{pmatrix}, \quad \sigma_3 = \begin{pmatrix} 1 & 0 \\ 0 & -1 \end{pmatrix}. \tag{A.19}$$

The components of $\vec{\sigma}$ satisfy the following commutation relation:

$$[\sigma^j, \sigma^k] = i\epsilon_{jkl}\sigma^l, \tag{A.20}$$

where ϵ_{jkl} is a completely antisymmetric tensor, that is $\epsilon_{jkl} = 0$, if any two indices are equal; and $\epsilon_{jkl} = (-1)^m$, where m is the number of transpositions from (123) to (jkl). It is convenient to define the spin up and down operators, which are linear combinations of σ_1

and σ_2 operators:

$$\sigma_+ = \sigma_1 + i\sigma_2 = \begin{pmatrix} 0 & 0 \\ 1 & 0 \end{pmatrix},$$

$$\sigma_- = \sigma_1 - i\sigma_2 = \begin{pmatrix} 0 & 1 \\ 0 & 0 \end{pmatrix}. \tag{A.21}$$

A.2.2 *Helicity and spinors*

Helicity λ is defined as a projection of spin on the direction of motion. Particles with positive projection of spin on the direction of motion ($\lambda = +1/2$) are referred to as right-handed, and left-handed otherwise ($\lambda = -1/2$).

Left/right-handed helicity projection operators are given by

$$\Pi_L = \frac{1}{2}\left(1 - \frac{\vec{\sigma}\vec{p}}{|p|}\right), \quad \Pi_R = \frac{1}{2}\left(1 + \frac{\vec{\sigma}\vec{p}}{|p|}\right). \tag{A.22}$$

The two helicities form a complete set of eigenvalues of the spin operator for a particle with $s = 1/2$. The corresponding eigenvectors, referred to as spinors, are

$$\chi_+ = \begin{pmatrix} 1 \\ 0 \end{pmatrix}, \quad \chi_- = \begin{pmatrix} 0 \\ 1 \end{pmatrix},$$

which correspond to spin up, spin down states, respectively. It is obvious that spin up operator turns spin down state into spin up state and vice versa:

$$\sigma_+\chi_- = \chi_+, \quad \sigma_-\chi_+ = \chi_-. \tag{A.23}$$

A general spinor has a form

$$\chi = \begin{pmatrix} \cos\frac{\theta}{2} \\ e^{i\phi}\sin\frac{\theta}{2} \end{pmatrix} = \cos\frac{\theta}{2}\chi_+ + e^{i\phi}\sin\frac{\theta}{2}\chi_-, \tag{A.24}$$

where θ is the polar angle and ϕ is the azimuthal angle of the spin vector.

A.2.3 *Dirac equation and Dirac spinors*

In the relativistic case, a free particle with mass m and a spin of $1/2$ is described by a wave function ψ, which is a solution to the Dirac's equation of the following form:

$$(i\gamma_\mu \partial^\mu - m)\psi = 0. \tag{A.25}$$

The 4×4 γ matrices in Dirac–Pauli representation are as follows:

$$\gamma_0 = \begin{pmatrix} I & 0 \\ 0 & -I \end{pmatrix}, \quad \gamma_i = \begin{pmatrix} 0 & \sigma_i \\ -\sigma_i & 0 \end{pmatrix}. \tag{A.26}$$

where I is a 2×2 identity matrix and σ_i are Pauli matrices. We note that this equation can be derived from Eq. (2.65) following the Euler–Lagrange formalism, though historically Dirac derived it from different principles. The solution to Dirac equation is

$$\psi = \sqrt{E + m}\, u(p)\, e^{-ipx}, \tag{A.27}$$

where $u^{(s)}(p)$ is a four-dimensional Dirac spinor:

$$u = \begin{pmatrix} \chi \\ \frac{\vec{\sigma}\vec{p}}{E+m}\chi \end{pmatrix},$$

with χ being a Pauli spinor given by Eq. (A.24).

Spinors must satisfy the completeness relation, that is the sum over helicities is

$$\sum_{\lambda=+,-} u^\lambda(p)\bar{u}^\lambda(p) = \gamma_\mu p^\mu + m \tag{A.28}$$

For compactness of notation, we sometimes label a spinor corresponding to an electron by $e(p)$, neutrino by $\nu(p)$, etc.

A.2.4 Axial and vector properties of γ matrices

The four contravariant gamma or Dirac matrices are defined (in the Dirac representation) as follows:

$$\gamma^0 = \begin{pmatrix} 1 & 0 & 0 & 0 \\ 0 & 1 & 0 & 0 \\ 0 & 0 & -1 & 0 \\ 0 & 0 & 0 & -1 \end{pmatrix}, \quad \gamma^1 = \begin{pmatrix} 0 & 0 & 0 & 1 \\ 0 & 0 & 1 & 0 \\ 0 & -1 & 0 & 0 \\ -1 & 0 & 0 & 0 \end{pmatrix},$$

$$\gamma^2 = \begin{pmatrix} 0 & 0 & 0 & -i \\ 0 & 0 & i & 0 \\ 0 & i & 0 & 0 \\ -i & 0 & 0 & 0 \end{pmatrix}, \quad \gamma^3 = \begin{pmatrix} 0 & 0 & 1 & 0 \\ 0 & 0 & 0 & -1 \\ -1 & 0 & 0 & 0 \\ 0 & 1 & 0 & 0 \end{pmatrix}. \quad \text{(A.29)}$$

γ^0 is the time-like, hermitian matrix. The other three are spacelike, antihermitian matrices. The defining property for the gamma matrices is the anticommutation relation

$$\{\gamma^\mu, \gamma^\nu\} = \gamma^\mu \gamma^\nu + \gamma^\nu \gamma^\mu = 2g^{\mu\nu} I_4, \quad \text{(A.30)}$$

where $\{,\}$ is the anticommutator, $g^{\mu\nu}$ is the Minkowski metric, and I_4 is the 4×4 identity matrix.

Covariant gamma matrices are defined by $\gamma_\mu = g_{\mu\nu}\gamma^\nu = \{\gamma^0, -\gamma^1, -\gamma^2, -\gamma^3\}$.

It is useful to define a product of the four gamma matrices as $\gamma^5 = \sigma_1 \otimes I$, so that

$$\gamma^5 := i\gamma^0 \gamma^1 \gamma^2 \gamma^3 = \begin{pmatrix} 0 & 0 & 1 & 0 \\ 0 & 0 & 0 & 1 \\ 1 & 0 & 0 & 0 \\ 0 & 1 & 0 & 0 \end{pmatrix}. \quad \text{(A.31)}$$

A.2.5 Chirality operator

Left/right-handed chirality projection operators are given by

$$P_L = \frac{1 - \gamma_5}{2}, \quad P_R = \frac{1 + \gamma_5}{2}. \quad \text{(A.32)}$$

In the ultrarelativistic limit, chirality approximates helicity, which motivates the naming — left or right-handed. Unlike helicity, chirality has no simple visualization. This is a quantum number that is defined through its operator. Weak currents are defined based on the chiral structure.

A.3 Spin 1 Particles

A.3.1 *Photons*

The electromagnetic field is described by its potential in covariant form $A_\mu = (\phi, \vec{A})$. The electromagnetic tensor $F_{\mu\nu}$ contains all the components of the electric and magnetic fields, as follows:

$$F^{\mu\nu} = \partial^\mu A^\nu - \partial^\nu A^\mu = \begin{pmatrix} 0 & -E_x & -E_y & -E_z \\ E_x & 0 & -B_z & B_y \\ E_y & B_z & 0 & -B_x \\ E_z & -B_y & B_x & 0 \end{pmatrix}. \qquad (A.33)$$

A free photon is described by

$$\partial_\nu \partial^\nu A_\mu = 0, \qquad (A.34)$$

and the solution to this equation has the form

$$A_\mu = \epsilon_\mu(p) e^{-ipx}, \qquad (A.35)$$

where p is photon's four-momentum and $\epsilon_\mu(p)$ is its four-dimensional polarization vector. Plugging Eq. (A.35) into Eq. (A.34), we get $p^2 = 0$, meaning that photon is massless. In the absence of currents, the following is true:

$$0 = \partial_\mu A^\mu = -i\epsilon_\mu p^\mu e^{-ipx}. \qquad (A.36)$$

Hence,

$$\epsilon_\mu p^\mu = 0. \qquad (A.37)$$

which means that ϵ_μ has only two independent components, both transverse to the photon's direction of motion, i.e., there are two degrees of freedom that describe the photon's polarization. Choosing the z axis to be directed along the photon's momentum, we conclude that the polarization vector lies in the (x, y) plane and can be presented as a linear combination of two transversely polarized states:

$$\epsilon_1 = (0, 1, 0, 0), \quad \epsilon_2 = (0, 0, 1, 0), \tag{A.38}$$

or alternatively two circularly polarized states:

$$\epsilon_- = \frac{1}{\sqrt{2}}(0, 1, -i, 0), \quad \epsilon_+ = -\frac{1}{\sqrt{2}}(0, 1, i, 0), \tag{A.39}$$

corresponding to -1 and $+1$ values of the spin projection on the z axis, respectively.

A.3.2   *Massive vector bosons*

For a spin-1 particle V with mass m_V, we have

$$(\partial_\nu \partial^\nu + m_V^2)V_\mu = 0. \tag{A.40}$$

Seeking the solution in the form

$$V_\mu = \epsilon_\mu(p)e^{-ipx}, \tag{A.41}$$

we get the condition $p^2 = m_V^2$, which is the dispersion equation for a massive particle. In this case, the condition

$$\epsilon_\mu p^\mu = 0, \tag{A.42}$$

reduces the number of degrees of freedom from four to three. The transverse polarization vectors are described by Eq. (A.38). Seeking the solution for the longitudinal polarization in the form

$$\epsilon_L = \frac{1}{\sqrt{\alpha^2 + \beta^2}}(\alpha, 0, 0, \beta), \tag{A.43}$$

we get

$$\alpha E - \beta p_z = 0. \tag{A.44}$$

Hence,

$$\epsilon_L = \frac{1}{m}(p_z, 0, 0, E). \tag{A.45}$$

The completeness relationship for gauge boson states has the following form:

$$\sum_\lambda \epsilon_\mu \epsilon_\nu^* = g_{\mu\nu} - \frac{p_\mu p_\nu}{m_V^2}. \tag{A.46}$$

For massless photons, the second term is absent.

A.4 Some Basics on Lagrangians

The Lagrangian function is simply the combination of the kinetic and potential energies with a minus sign: $L = T - U$. This function can be used to derive the equation of motion of the physical system, by means of the Euler–Lagrange formula:

$$\frac{d}{dt}\left(\frac{\partial L}{\partial \dot{q}_i}\right) - \frac{\partial L}{\partial q_i} = 0, \tag{A.47}$$

where q and t are the generalized coordinates, and $\dot{q}$ is the time derivative of that coordinate. In field theory, the Lagrangian $L(q_i, \dot{q}_i)$ as a function of generalized coordinates is replaced by the Lagrangian density $\mathcal{L}(\varphi_i, \partial_\mu \varphi_i)$, which is a function of the fields φ_i, their derivatives, and possibly the space and time coordinates themselves. In this case, the Euler–Lagrange equations describe the geodesic flow of the field φ as a function of time:

$$\frac{\partial \mathcal{L}}{\partial \varphi} - \partial_\mu \left(\frac{\partial \mathcal{L}}{\partial(\partial_\mu \varphi)}\right) = 0. \tag{A.48}$$

For example, the Lagrangian for a free real scalar field ϕ:

$$\mathcal{L}_{\text{scalar}} = \frac{1}{2}(\partial_\mu \phi)(\partial^\mu \phi) - \frac{1}{2}m^2\phi^2 \xrightarrow{\text{Euler–Lagrange}} \underbrace{(\partial_\mu \partial^\mu + m^2)\psi = 0}_{\text{Klein–Gordon equation}}. \tag{A.49}$$

And, similarly, for free fermions described by the wave function ψ:

$$\mathcal{L}_{\text{fermion}} = i\bar{\psi}\gamma_\mu\partial^\mu\psi - m\bar{\psi}\psi \xrightarrow{\text{Euler–Lagrange}} \underbrace{(i\gamma_\mu\partial^\mu - m)\psi = 0}_{\text{Dirac equation}}.$$

$$(\text{A.50})$$

In general, the Lagrangian for a real scalar particle will contain the following terms:

$$\mathcal{L} = \underbrace{(\partial_\mu\phi)^2}_{\text{kinetic term}} + \underbrace{C}_{\text{constant}} + \underbrace{\alpha\phi}_{?} + \underbrace{\beta\phi^2}_{\text{mass term}} + \underbrace{\gamma\phi^3}_{\text{3-point int.}}$$

$$+ \underbrace{\delta\phi^4}_{\text{4-point int.}} + \cdots \qquad (\text{A.51})$$

The constant term is never relevant because it does not appear in the equations of motion and the linear term in the field should not be present because it has no direct interpretation. The quadratic term represents the mass of the particle and the triple and quartic terms describe self-interaction terms of the field with itself.

A.4.1 *Noether's theorem*

The theorem can simply be stated as follows: For each symmetry of the Lagrangian, there is a conserved quantity. We test this "symmetry" by changing the coordinates by a small amount and checking the Lagrangian has no first-order change in these quantities. The "conserved quantity" will be a quantity that does not change in time.

Proof: Let's change the coordinates by a small amount ϵ and a function that may depend on other coordinates:

$$q_i \longrightarrow q_i + \epsilon K_i(q), \qquad (\text{A.52})$$

where each $K_i(q)$ may be a function of all the q_i, which we can denote by the shorthand, q. If we require that the Lagrangian is invariant under this change, then

$$0 = \frac{d\mathcal{L}}{d\epsilon} = \sum_i \left(\frac{\partial \mathcal{L}}{\partial q_i} \frac{\partial q_i}{\partial \epsilon} + \frac{\partial \mathcal{L}}{\partial \dot{q}_i} \frac{\partial \dot{q}_i}{\partial \epsilon} \right) = \sum_i \left(\frac{\partial \mathcal{L}}{\partial q_i} K_i + \frac{\partial \mathcal{L}}{\partial \dot{q}_i} \dot{K}_i \right).$$

$$(A.53)$$

We now use the Euler–Lagrange equation (Eq. (A.47)) to rewrite the first term in the sum as follows:

$$0 = \sum_i \left(\frac{d}{dt} \left(\frac{\partial \mathcal{L}}{\partial \dot{q}_i} \right) K_i + \frac{\partial \mathcal{L}}{\partial \dot{q}_i} \dot{K}_i \right) = \frac{d}{dt} \left(\sum_i \frac{\partial \mathcal{L}}{\partial \dot{q}_i} K_i \right), \quad (A.54)$$

which means that the quantity

$$p(q, \dot{q}) \equiv \sum_i \frac{\partial \mathcal{L}}{\partial \dot{q}_i} K_i(q), \qquad (A.55)$$

does not change with time. This quantity is generally called the *conserved momentum* (but does not have to have the units of linear momentum). In field theory, we use the term *conserved charge*.

The simplest case would be a Lagrangian that does not depend on one of the coordinates (for example q_k), and it is therefore invariant (symmetric) under any change: $q_k \to q_k + \epsilon$. This is a translational invariance. The conserved quantity in this case is the linear momentum p_k:

$$p_k = \frac{\partial \mathcal{L}}{\partial \dot{q}_k}. \qquad (A.56)$$

Think of the two-dimensional movement of an object under gravity:

$$\mathcal{L} = \frac{1}{2} m(\dot{x}^2 + \dot{y}^2) - U(y), \qquad (A.57)$$

where the uniform potential energy $U(y) = mgy$ is independent of x. This Lagrangian clearly does not depend on x, or put another way, there is no force along x: $F_x = \partial \mathcal{L}/\partial x = 0$, and therefore, the horizontal component of the linear momentum $p_x = \partial \mathcal{L}/\partial \dot{x} = m\dot{x}$ will be conserved.

A.5 Electromagnetic Interactions

A.5.1 *Currents and electric charge*

A conserved current corresponding to the solution of the Dirac equation as shown in Eq. (A.27) is given by

$$J_\mu^{\text{EM}} = \bar{u}(p)\gamma_\mu \hat{Q} u(p) \,, \tag{A.58}$$

where the eigenvalue of the operator $\hat{Q}$ is the electric charge of the particle:

$$Q = \int J_0^{\text{EM}} d^3x. \tag{A.59}$$

A.5.2 *QED Lagrangian*

The full QED Lagrangian has a form of

$$\mathcal{L}_{\text{QED}} = \bar{\psi}(i\gamma_\mu \partial_\mu - m)\psi + \underbrace{e\bar{\psi}\gamma_\mu A_\mu \psi - \frac{1}{4}F_{\mu\nu}F_{\mu\nu}}_{\mathcal{L}_{\text{EM}}}. \tag{A.60}$$

The first term describes a free fermion. We saw in Eq. (A.50) how that term produces the Dirac equation of motion. The second term represents the interaction of the electromagnetic field with the current ($J_\mu = e\bar{\psi}\gamma_\mu \psi$). The last term presents the kinetic energy of the electromagnetic field. If we apply the Euler–Lagrange equation (Eq. (A.48)) to the second and third terms (that represent an electromagnetic field with sources, $\mathcal{L}_{\text{EM}}$), we get the Maxwell equations, as follows:

$$\frac{\partial \mathcal{L}_{\text{EM}}}{\partial A_\nu} - \partial_\mu \left(\frac{\partial \mathcal{L}_{\text{EM}}}{\partial(\partial_\mu A_\nu)} \right) = 0 \longrightarrow \partial_\mu F^{\mu\nu} = J^{\text{EM}\nu}. \tag{A.61}$$

A.6 Summary of Electroweak Interactions

A.6.1 *Coupling constants*

The strength of the coupling in the SU(2) weak isospin group is g; in hypercharge U(1) group, it is g'; in QED, it is e. Mediators of hypercharge field and the neutral component of W triplet mix to form the observed electromagnetic current and the neutral weak interaction. The angle of mixing is θ_W. The following relationships between these constants are true:

$$g \sin \theta_W = e = g' \cos \theta_W.$$

From this, we can derive

$$\sin \theta_W = e/g, \quad \cos \theta_W = e/g'.$$

These constants can be related to the Fermi constant describing the four-fermion vertices through the following equation:

$$\frac{G_F}{\sqrt{2}} = \frac{g^2}{8m_W^2}. \tag{A.62}$$

A.6.2 *Masses*

The W boson mass is

$$m_W = \frac{g}{2}v. \tag{A.63}$$

The Z boson mass is

$$m_Z = \frac{g}{2 \cos \theta_W}v. \tag{A.64}$$

Fermion masses are given by

$$m_f = \frac{g_f}{\sqrt{2}}v. \tag{A.65}$$

The Higgs boson mass is related to the self-interaction constant λ and the vev:

$$m_h = v\sqrt{2\lambda}. \tag{A.66}$$

Suggested Reading for Appendix A

[1] Donald Hill Perkins. *Introduction to high energy physics; 4th ed.* Cambridge: Cambridge Univ. Press, 2000. URL: https://cds.cern.ch/record/396126.

[2] Giovanni Punzi. "Sensitivity of searches for new signals and its optimization". *eConf* C030908 (2003), MODT002. arXiv: physics/0308063.

[3] Glen Cowan et al. "Asymptotic formulae for likelihood-based tests of new physics". *Eur. Phys. J. C* 71 (2011). [Erratum: Eur.Phys.J.C 73, 2501 (2013)], p. 1554. arXiv: 1007.1727 [physics.data-an].

[4] LHC Higgs Cross Section Working Group Collaboration. "Handbook of LHC Higgs Cross Sections: 4. Deciphering the Nature of the Higgs Sector". *CERN Yellow Report* 2/2017 (Oct. 2016). arXiv: 1610.07922 [hep-ph].

List of Figures with Credits

Index